Defending APIs

Uncover advanced defense techniques to craft secure application programming interfaces

Colin Domoney

Defending APIs

Copyright © 2024 Packt Publishing

Group Product Manager: Pavan Ramchandani

Publishing Product Manager: Prachi Sawant

Book Project Manager: Ashwin Dinesh Kharwa

Senior Editor: Isha Singh

Technical Editor: Nithik Cheruvakodan

Copy Editor: Safis Editing

Proofreader: Safis Editing

Indexer: Tejal Daruwale Soni

Production Designer: Vijay Kamble

DevRel Marketing Coordinator: Marylou De Mello

First published: February, 2024

Production reference: 112012024

Published by Packt Publishing Ltd.

Grosvenor House

11 St Paul's Square

Birmingham

B3 1RB, UK

ISBN 978-1-80461-712-0

www.packtpub.com

To everyone tasked with developing or securing software: you have an awesome responsibility, and this book is dedicated to making that just a little bit easier.

For my late father, Athol, who from my earliest memory encouraged me to apply myself fully to my learning and education, and would be very proud of what I have achieved.

Colin Domoney

Foreword

In a digital world completely reliant on APIs, mastering their security is not just a necessity, it's an imperative. Every organization is dependent on internet technology to function and that technology is dependent on APIs that were created by the developers at those organizations or the developers at their technology suppliers. Yet, these developers and the security professionals who need to work with them to manage this increasingly important risk lack knowledge on how to effectively secure APIs. Daily, we see the downfall of this missing knowledge. Data breaches and ransomware are endemic. One of the biggest causes is the vulnerable APIs that expose an organization's data, network, and transactions to the internet, and attackers!

It's my privilege to introduce Colin Domoney's *Defending APIs*," a book that stands as a testament to the critical importance of securing APIs in today's interconnected landscape. Colin, with his vast experience in software development and security, delves deep into the world of APIs. His journey through the intricacies of API vulnerabilities, attack strategies, and robust defense mechanisms is both enlightening and essential. This book is not just about theories; it's a reflection of real-world scenarios, offering practical insights and solutions.

Understanding the fundamentals of how a technology is designed and implemented is critical to understanding how to secure it. *Part One* lays a foundational understanding of API security, addressing its holistic nature and dissecting core components such as protocols and data formats. It also dives into the OWASP API Security Top 10, providing insights into common vulnerabilities and high-profile breaches.

It is necessary to understand how common failures in secure design and implementation are taken advantage of by attackers. *Part Two* focuses on attacking APIs, equipping readers with knowledge of adversarial techniques and tools, crucial for crafting robust defenses.

With an understanding of the technical mechanics of APIs and how attackers operate in the real world, you will be ready for the final segment. In *Part Three*, Colin imparts an in-depth guide on defending APIs, emphasizing tools and techniques across the development lifecycle and exploring the future of API security in microservices architecture.

As someone who has dedicated a career to advocating for secure software design, I find Colin's approach in this book particularly compelling. He bridges the gap between understanding the problem and implementing effective solutions, making this book a valuable resource for both security professionals and software developers alike.

Defending APIs is a journey through the complexities of API security, guided by a seasoned expert. Colin's work here is not just informative; it's inspiring, pushing us to think differently about how we protect this important glue that holds together our shared digital infrastructure.

In a time where digital security is paramount, this book serves as a beacon, guiding the way toward a more secure digital future. I highly recommend it to anyone who's serious about understanding and implementing API security in their organization.

Chris Wysopal

Veracode co-founder and CTO

December 31, 2023

In the famous words of Marc Andreessen, software is eating the world. In that world, APIs are the entry door to absolutely everything, from your bank to your brand-new washing machine, moving data across a wide variety of systems. Data is the new gold, and once you know that, you understand why APIs are the target of so many cyber-attacks.

When we started 42Crunch, we had to convince people APIs were a new attack target and, more importantly, a different attack target. Architectures based on APIs are different and, as such, you need new ways to protect them. The traditional protection measures at the edge of the network and vulnerability detection via code analysis at the end of the development cycle simply won't work and won't scale for API-based applications.

Fast-forward five years and API security is now one of the hottest topics in IT, with every enterprise realizing they need to work diligently to secure their API development lifecycle.

Colin has helped numerous 42Crunch customers build a strategy toward developing rugged APIs and authored our API security maturity model. His experience, looking at the issue both from an attacker and defender perspective, is condensed into his book. Should you be a CISO building a new security program for APIs or a developer trying to understand how to build better, resilient APIs, this book is for you.

I wish you a great journey securing your APIs, with *Defending APIs* by your side!

Isabelle Mauny

42Crunch co-founder and CTO

January 2, 2024

Contributors

About the author

Colin Domoney has a long and varied career both in development and, more recently, as an advocate for secure software systems. His career started with developing mission-critical military and medical systems in South Africa, where he learned the importance of developing systems that are rugged and robust. He has applied this experience across multiple industries, including automotive, financial services, consumer electronics, and others.

Most recently, he has applied his skills to building large-scale application security programs for some of the largest organizations globally, helping them scale their software security capabilities. As an independent security consultant, he specializes in coaching and teaching developers how best to build secure software. He is the chief technology evangelist at 42Crunch, focused on advocating for secure API development best practices and developing APIs that are secure by design. He is an avid learner and technologist at heart and is happiest discovering new technologies or approaches to securing software.

In addition to his advocacy role, where he is a regular conference presenter and webinar host, Colin is also the curator of the *APISecurity.io* newsletter, featuring the latest in API security news. It was through this newsletter that he found a passion for API security in particular, leading to the journey that has led to the book you have in front of you now. He is currently working on a course and workshop to accompany the book.

About the reviewer

Peter Chestna has been a software engineer for over 30 years. For more than half his career, he has specialized in application security, commonly called AppSec, from both the vendor and practitioner side. Starting as a developer at Veracode in 2006, he rose through the ranks to lead the SaaS engineering delivery team. He left to transform and lead the AppSec program at Bank of Montreal (BMO) for several years before returning to the vendor side to become the CISO of North America at Checkmarx.

A lifelong learner, he devotes significant time to reading about psychology and leadership, software development practices, and cybersecurity. He speaks internationally at Agile, DevOps, and security conferences, with a particular affinity for DevOpsDays. Pete has been granted three patents.

Table of Contents

3

Understanding Common API Vulnerabilities   53

4

Investigating Recent Breaches   75

Part 2: Attacking APIs

5

Foundations of Attacking APIs 105

6

Discovering APIs 131

Part 3: Defending APIs

9

Defending against Common Vulnerabilities 215

10

Securing Your Frameworks and Languages 241

Preface

In today's hyper-connected digital world, APIs are ubiquitous in providing the connecting tissue between systems and services. The growth of APIs continues at an exponential pace, with almost every developer either responsible for creating APIs of their own or consuming APIs as part of their solution. Unfortunately, attackers have shifted their focus to attacking APIs first, and the alarming rise in API incidents and breaches is testimony to the challenges developers face in producing APIs that are secure and robust.

This book is intended to be the primary reference for developers wishing to build secure APIs, and for security teams wanting to get a grip on the unique challenges of securing APIs. The book has a strong practical focus, with extensive code samples and tooling and references to real-world API breaches. The first part covers the basics of APIs and security, including common API vulnerabilities.

As a defender, it is essential to understand the methods of your adversaries, and I will guide you through the common skills and techniques employed by attackers. This will equip you with the skills to thoroughly test your APIs for common API weaknesses and vulnerabilities. The book addresses the full spectrum of API security, from pre-emptive secure-by-design practices as part of a shift-left approach to state-of-the-art runtime protection as part of a shield-right approach.

Finally, I draw on my experience of building large-scale software security programs to provide you with insights and strategies to establish and then mature your own API security programs.

Join me on this exciting journey into the world of API security and ensure that your APIs are never a point of weakness for attackers.

Who this book is for

This book is a perfect companion for security professionals responsible for API security. For AppSec teams, there is a focus on security tooling and integration and guidance on how to build an AppSec program targeting API security. For SecOps teams, there is in-depth coverage of API protection and monitoring to protect APIs at runtime.

The book is intended to be a reference for API developers, helping them understand the threats and attacks their APIs are likely to face and how to defend against the most common attack types, with a focus on API design first to enable shift-left for API security.

Finally, the book will appeal to system architects needing to understand best practices for secure API design and implementation.

What this book covers

Chapter 1, What Is API Security?, provides an introduction to the topic of API security and why it is important and distinct from web application security. This chapter also provides an understanding of the basics of APIs and their data formats, covering the key elements of API security and goals.

Chapter 2, Understanding APIs, covers the fundamentals of the HTTP protocol and the different types of APIs currently in use. The key topics of authentication and authorization are covered, along with the use of tokens and keys.

Chapter 3, Understanding Common API Vulnerabilities, provides in-depth coverage of the OWASP API Security Top 10 vulnerabilities (both the 2019 and 2023 variants), how vulnerabilities differ from abuse cases, and how APIs can expose business logic vulnerabilities.

Chapter 4, Investigating Recent Breaches, is an eye-opening look at some of the most significant API security breaches in the last few years, where we examine what went wrong and how such vulnerabilities could be prevented in the future.

Chapter 5, Foundations of Attacking APIs, provides a foundation of how adversaries attack APIs, including their methods, the common tools, and the skills they utilize.

Chapter 6, Discovering APIs, illuminates the various passive and active methods used by adversaries to discover APIs. We will also examine reconnaissance methods used to understand implementations and to evade common defense methods.

Chapter 7, Attacking APIs, provides hands-on guidance on how to attack APIs, focusing on the following areas – authentication and authorization attacks, data-based attacks, injection attacks, and other common attack types.

Chapter 8, Shift-Left for API Security, focuses on core activities that can be used to shift API security left, including leveraging the OpenAPI specification and the positive security model, how to threat model APIs, and the automation of API security within CI/CD pipelines.

Chapter 9, Defending against Common Vulnerabilities, covers the core topics of the defensive patterns and techniques that can be used to defend APIs against the following vulnerabilities – authentication and authorization vulnerabilities, data vulnerabilities, and implementation vulnerabilities.

Chapter 10, Securing Your Frameworks and Languages, moves the focus onto securing languages and frameworks using a "design-first" approach, including the use of code generation tooling, OpenAPI generation with popular frameworks, and patterns to secure these frameworks.

Chapter 11, Shield-Right for APIs with Runtime Protection, emphasizes the critical role played by so-called "shield-right" techniques, including secure and hardened environments, WAFs for API protection, the use of API gateways and management portals, API firewalls and, finally, monitoring APIs at runtime.

Chapter 12, Securing Microservices, looks at the exciting world of APIs within a microservices architecture, where we learn to apply our existing knowledge in a microservices landscape, focusing on securing the foundations, connectivity, and access control.

Chapter 13, Implementing an API Security Strategy, concludes our journey into API security by providing focused guidance on how to build an API security strategy, including the selection of a roadmap and KPIs, and how to plan and execute your strategy.

To get the most out of this book

The following are the software/hardware covered in the book and OS requirements:

Software/hardware covered in the book	Operating system requirements
Angular 9	Windows, macOS, or Linux
TypeScript 3.7	
ECMAScript 11	

If you are using the digital version of this book, we advise you to type the code yourself or access the code from the book's GitHub repository (a link is available in the next section). Doing so will help you avoid any potential errors related to the copying and pasting of code.

Download the example code files

You can download the example code files for this book from GitHub at `https://github.com/PacktPublishing/Defending-APIs`. If there's an update to the code, it will be updated in the GitHub repository.

We also have other code bundles from our rich catalog of books and videos available at `https://github.com/PacktPublishing/`. Check them out!

Conventions used

There are a number of text conventions used throughout this book.

`Code in text`: Indicates code words in text, database table names, folder names, filenames, file extensions, pathnames, dummy URLs, user input, and Twitter handles. Here is an example: "In this example, we can see that a combination of `hellopixi` and `user@acme.com` resulted in a `200 OK` status code and the return of a JWT."

A block of code is set as follows:

```
<note>
    <to>Colin</to>
    <priority>High</priority>
    <heading>Reminder</heading>
    <body>Learn about API security</body>
</note>
```

When we wish to draw your attention to a particular part of a code block, the relevant lines or items are set in bold:

```
{
    "openapi": "3.0.0",
    "info": {
        "version": "1.0.0",
        "title": "Swagger Petstore",
        "license": {
            "name": "MIT" }
```

Any command-line input or output is written as follows:

```
colind@mbm: ~ # sudo nmap -sn 192.168.9.0/24
```

Bold: Indicates a new term, an important word. For instance, words in menus or dialog boxes appear in **bold**. Here is an example: "**YAML Ain't Markup Language (YAML)** is another common internet format, similar to JSON in its design goals".

> **Tips or important notes**
> Appear like this.

Get in touch

Feedback from our readers is always welcome.

General feedback: If you have questions about any aspect of this book, email us at customercare@packtpub.com and mention the book title in the subject of your message.

Errata: Although we have taken every care to ensure the accuracy of our content, mistakes do happen. If you have found a mistake in this book, we would be grateful if you would report this to us. Please visit www.packtpub.com/support/errata and fill in the form.

Piracy: If you come across any illegal copies of our works in any form on the internet, we would be grateful if you would provide us with the location address or website name. Please contact us at `copyright@packt.com` with a link to the material.

If you are interested in becoming an author: If there is a topic that you have expertise in and you are interested in either writing or contributing to a book, please visit `authors.packtpub.com`.

Share Your Thoughts

Once you've read *Defending APIs*, we'd love to hear your thoughts! Scan the QR code below to go straight to the Amazon review page for this book and share your feedback.

`https://packt.link/r/1804617121`

Your review is important to us and the tech community and will help us make sure we're delivering excellent quality content.

Download a free PDF copy of this book

Thanks for purchasing this book!

Do you like to read on the go but are unable to carry your print books everywhere?

Is your eBook purchase not compatible with the device of your choice?

Don't worry, now with every Packt book you get a DRM-free PDF version of that book at no cost.

Read anywhere, any place, on any device. Search, copy, and paste code from your favorite technical books directly into your application.

The perks don't stop there, you can get exclusive access to discounts, newsletters, and great free content in your inbox daily

Follow these simple steps to get the benefits:

1. Scan the QR code or visit the link below

https://packt.link/free-ebook/9781804617120

2. Submit your proof of purchase

3. That's it! We'll send your free PDF and other benefits to your email directly

Part 1:
Foundations of API Security

In this part, you will gain a foundational understanding of the key components of **Application Programming Interface (API)** security. You will understand the need for a holistic approach to API security as APIs become the main target for hackers. We then take a look at the core building blocks of APIs, including protocols, data formats, authentication, authorization, and their role in security. The OWASP API Security Top 10 takes you on a walk-through of the most common vulnerabilities, and finally, this part concludes with detailed post-mortems of some of the highest-profile breaches in recent times.

This part has the following chapters:

- *Chapter 1, What Is API Security?*
- *Chapter 2, Understanding APIs*
- *Chapter 3, Understanding Common API Vulnerabilities*
- *Chapter 4, Investigating Recent Breaches*

1

What Is API Security?

A frequently quoted view is that there is no such thing as API security; it's just an evolution of application security we have been practicing for the last two decades. However, I believe that it is a discrete and important discipline. Join me on the journey into API security.

APIs are the backbone of a modern digital economy, allowing the exchange of critical data and the interconnectivity of different systems. APIs are the fuel that have fired digital innovation for the last decade. Given the critical role of APIs in our digital world, it is vital that they are secure. This chapter sets out the foundational concepts of APIs, particularly in relation to security.

In this chapter, we will examine *exactly* what is meant by API security and understand the key elements of this exciting and emerging security domain, covering topics such as the following:

- The importance of API security
- Understanding the basics of APIs
- API data formats
- Elements of API security and API security goals

Why API security is important

The **Open Web Application Security Project** (**OWASP**) published its first OWASP API Security Top 10 list in December 2019, and since then, the API security community has grown rapidly, with API security start-ups attracting significant investment and increasing interest from developers and security practitioners alike to learn resources on the topic. Unfortunately, during this period, there has also been a marked rise in the number of security incidents relating to insecure APIs. Recent analysis suggests a 681% rise in API attacks and that nearly one in two organizations has experienced a security incident related to APIs.

In a way, *APIs are a victim of their own success* – because of their rapid proliferation and the high economic value of the data they protect, they are now the most popular target for attackers.

We'll now have a look at the so-called API economy and how the near-exponential growth in the number of APIs creates challenges for organizations as they become the favorite vector for attackers.

The growth of the API economy

To fully appreciate the importance of API security, let us first consider the growth of the so-called API economy. Let's understand a bit more about what is meant by this term – Forbes defines an API economy as "*an enabler for turning a business or organization into a platform.*" A platform can leverage APIs to do the following:

- Provide services and data to consumers for a price
- Consume services and data from other providers to enhance your business

What is the API economy?

The API revolution has led to the emergence of API-first businesses such as Twilio and has allowed other organizations to expose their core offerings via APIs (Google Maps is a good example). The disruptive nature of the API economy is best seen in the financial services industry – typically, this has been an industry resistant to innovation due to regulatory and compliance requirements. By using APIs to expose selected core services, banks can embrace new models without disrupting their core IT systems. By adopting open standards that can be certified – such as the **Open Banking API** – banks can achieve interoperability while ensuring transactional integrity. The online money transfer service Wise uses APIs to provide B2B and B2C services and offers **banking-as-a-service** (**BaaS**) to third parties by renting out their APIs.

Advantages of an API economy

There are several key benefits to an API economy:

- **Reduced time-to-market**: Organizations can use APIs to consume services from third parties rather than having to create those services themselves, resulting in faster development life cycles.

- **Drive value**: Organizations can expose new and innovative services using APIs and open new markets.

- **Competitive advantage**: By getting to market faster and using APIs to drive innovation, adopters can increase their competitive advantage.

- **Improved efficiency**: APIs allow IT teams to deliver immediate value by exposing APIs, rather than having to build and deploy mobile or web applications.

- **Security**: Mobile and web applications expose a vast attack surface to adversaries. By focusing development on APIs, this attack surface can be reduced and focus given to the hardening and security of these APIs.

API adoption allows organizations to deliver more value and functionality while simultaneously reducing cost and time to market.

The scale of the API economy

It is difficult to provide an accurate estimation of the scale of an API economy, and even if it was possible, this estimate would soon be invalidated due to the nearly exponential growth of the space. The API community at Nordic APIs (`https://nordicapis.com/20-impressive-api-economy-statistics/`) has produced a survey on the scale of the API economy; the following are some headline figures:

* Over **90%** of developers use APIs

* The popular API test platform Postman has over **46 million** API collections

* **83%** of all internet traffic belongs to APIs

* There are over **2 million** API GitHub repositories

* The API management market is valued at **$5.1 billion** in 2023

* **93%** of communication service providers use OpenAPI specifications

* **91%** of organizations have had an API security incident

On the back of a growing API economy, major capital investment has poured into the market for API tool vendors, management platforms, and security tools.

Challenges to an API economy

The rapid adoption of APIs brings with it several challenges in addition to the benefits. The first challenge is that of inventory — because APIs can be easily built and deployed and have a finite lifetime, organizations are struggling to keep track of their API inventory, resulting in shadow (hidden) and zombie (outdated) APIs.

The second challenge is that of governance — as APIs proliferate, organizations face challenges with governing the development and deployment process, ensuring that data and privacy requirements are met and that the API life cycle is managed from cradle to grave.

The biggest challenge, however, is that of security. As noted earlier in this section, APIs can reduce an organization's overall attack surface; however, this comes at the cost of a new security paradigm – APIs are a new attack surface, and the threats are different. In the next section, we'll explore these security challenges in more detail.

APIs are popular with developers

Developers love APIs — nearly all developers work with APIs and nearly all modern architectures are API-centric. While containerization has driven the breakdown of the monolith and the emergence of microservices, it is APIs that form the connecting tissue between these services.

The benefit of APIs to developers are numerous, including the following:

- They form an abstraction between services and allow encapsulation of functionality.

- They define a clear interface via an OpenAPI specification that serves as a contract for the API.

- They allow a truly polyglot environment where different APIs can be implemented in the most suitable programming language for the task at hand.

- They simplify data exchange as APIs generally use JSON, XML, or YAML.

- They facilitate ease of testing, using tools such as Postman or tools that can validate API functionality against the OpenAPI Specification.

- They propel ease of development. The API development ecosystem is rich with powerful tooling for the development and testing of APIs. Moreover, fully featured API frameworks exist for most modern programming languages.

These factors have fueled the *API-first* paradigm where applications are built in a bottom-up approach, starting with the APIs, then the business logic, and the **user interface** (**UI**) last.

APIs are increasingly popular with attackers

While APIs are undoubtedly popular with developers, they are even more popular with attackers. Gartner reports that APIs are the number one attack vector for cybercriminals in 2022, and barely a week goes by without an API breach or vulnerability being disclosed.

There are several key reasons why APIs are a favored attack target:

- **APIs are likely to be publicly accessible**: By their nature, APIs are intended to be interconnected with other systems, requiring them to be exposed on public networks. This facilitates easy discovery and attack by adversaries.

- **APIs are often well documented**: To aid easy adoption and integration, good APIs should be documented using tools such as the **Swagger UI**. Unfortunately, such documentation can also be invaluable to attackers in understanding how the APIs work.

- **API attacks can be automated**: API interaction is *headless* (not requiring a UI or human interaction) and can easily be automated with scripts or dedicated attack tools. APIs are, in many cases, easier to attack than mobile or web applications.

- **APIs expose valuable data**: Most importantly, APIs are designed to allow access to key data assets (PII, financial, or market data), which are likely to be the highest prize for an attacker. Attackers increasingly attack APIs that inadvertently expose excessive data or allow mass exfiltration, which might not be the case with a well-crafted UI.

Your existing tools do not work well for APIs

The relatively recent emergence of APIs as the de facto conduit for application connectivity poses significant challenges to security teams and testers. Much of the existing **application security (AppSec)** tooling that exists was designed in an era when web applications were the primary asset to be protected. Common security tools such as **static application security testing (SAST)**, **dynamic application security testing (DAST)**, and **software composition analysis (SCA)** are far less effective in assessing APIs than they are with web or mobile applications.

Traditional perimeter protections such as network firewalls or **web application firewalls (WAFs)** are ineffective in protecting APIs, since they lack the context of the API interface and the expected request and response traffic. Such tools tend to be high in both false positives and false negatives.

More modern API technologies, such as **API management portals (APIMs)** and gateways, are essential for the operation of APIs at scale, but while they do provide security features, they do not address all attack vectors.

The key takeaway is that while tools are important as part of a defense strategy, they need to be augmented by solid defensive design and coding techniques — this is the focus of the final section of this book.

Developers often lack an understanding of API security

It is important to understand why insecure code exists in the first place if we want to address the problem.

Developers are, by nature, creative problem solvers who thrive on a challenge – unfortunately, this leads them to be *over-optimistic*, which can lead them to take shortcuts and optimizations, or perhaps work to unrealistic delivery schedules. This is so-called *happy path* coding, where developers do not fully appreciate how their code could fail or be misused by an attacker, sometimes with dire consequences.

Coupled with over-optimism is a sense of *over-confidence* – developers will assume they fully understand a problem but may be unaware that they are missing some crucial detail or subtlety, which again can have adverse effects. An example is the adoption of a new API framework and not carefully considering the default settings and deploying a vulnerable product.

Developers will often have a misplaced sense that bad things only happen to other people and not them. Despite witnessing examples of well-known breaches, many developers believe they will never fall victim to a similar misfortune. This general phenomenon is known as the schadenfreude effect.

The development process can be stressful with constant pressure to deliver to schedule, and this can result in compromising full implementation in favor of meeting deadlines. For example, this can include the omission of error handling code or data validation with the intent of coming back to implement them in later releases. With time pressures, this rarely happens, and code is often left in an incomplete state.

Often, developers inherit a code base to maintain that may contain significant *technical debt or legacy code*. Without a full understanding of the system and its complexities and foibles, developers may be disinclined to make changes to the code base in case they break functionality.

Exploring API building blocks

Before we can understand how to secure APIs, we need to dive into the building blocks of APIs. This section will cover the somewhat challenging topics of cryptography, hashing and signatures, encoding, and transport layer security. We will not go into a lot of detail, but it is important to grasp these basics.

Rate limiting

Public APIs are exposed to the internet and can easily be discovered by adversaries. One of the simplest attacks against an API is a **denial-of-service** (**DoS**) attack, in which automation is used to repeatedly and persistently attempt to access an API. Sustained DoS attacks can lead to the exhaustion of server resources, leading to a failure of the API or, most commonly, denying legitimate access to the API.

Brute-force attacks can also be used in **account takeover** (**ATO**) attacks, where either a sign-up endpoint or a password reset endpoint are flooded in attempts to guess passwords or hashes, using a **dictionary attack** (where a list of commonly used passwords is used).

Both types of attacks can be mitigated using rate-limiting technology, which limits repeated and frequent access from a particular IP address to a given endpoint. Rate-limiting applies a timeout window on the transactions and will return a `429 Too Many Requests` error.

Cryptography

Cryptography is a foundational element in securing data electronically – most simply, it is a mathematical transformation applied to data. Typically, *cleartext* (unencrypted) is transformed into *cyphertext* (encrypted), using an algorithm and a key. The cyphertext is no longer recognizable as the original cleartext and cannot be reverse-engineered to reveal the original cleartext without using the inverse transformation (decrypted), using the same algorithm and the key.

The choice of algorithm depends on the application; two broad types of algorithms are used:

- **Symmetric algorithm**: In this type, the same key is used to encrypt and decrypt data. The benefit of symmetric ciphers is that they are fast and safe; however, they pose a challenge in terms of the distribution of the shared key. Common symmetric key algorithms are DES, AES, and IDEA.

- **Asymmetric algorithm**: In this type, different keys are used to encrypt (using the public key) and decrypt (using the private key) data. Common asymmetric key algorithms are DDS, RSA, and ElGamal.

A fundamental challenge with cryptography is the exchange (and management) of *keys* between both parties. To this end, robust key-exchange protocols have been developed to securely exchange keys that prevent an eavesdropper from accessing keys in transit. The Diffie-Hellman exchange is the most used protocol.

Cryptography provides the following benefits:

- **Authentication**: By using *public-key cryptography*, it is possible to verify the identity of the originating party by using their public key to confirm they signed a message with their private key, which only they can access. By using *certificates*, it is possible to verify the validity (or trust) associated with public keys – this is the foundation of **Transport Layer Security** (**TLS**).

- **Nonrepudiation**: Using cryptography principles, transactions or documents can be audited to verify which parties had access to the resources. This prevents a receiving party from denying receipt; typically, this is used for bank transactions or document signatures.

- **Confidentiality**: The most obvious advantage of cryptography is to ensure that data is kept private, both in transit and at rest in storage. Only persons in possession of a valid key can decrypt and access the data.

- **Integrity**: Finally, cryptography can be used to ensure the integrity of data, verifying that it has not been modified in transit. By transmitting a *fingerprint* of the data along with it, the receiver can verify that the received data is the same that was transmitted by re-calculating the fingerprint and validating it against the one received.

Hashes, HMACs, and signatures

An important application of cryptography principles relates to ensuring the integrity of messages in transit.

Hashes are the most elementary technique, in which a block of data is passed through an algorithm to produce a *digest* of the data; typically this is a fixed-length string much shorter than the input data. Common hashing algorithms include SHA2 and MD5. Key properties of hash functions are that they are *one-way functions* or irreversible (the input cannot be obtained from the digest, and the digests are unique so that no two blocks of data will produce the same digest). Hashes are used to verify the integrity of data.

Hashed Authentication Message Codes (**HMACs**) are similar to hashes in that they produce a digest that is then encrypted with a symmetric algorithm and passed to the recipient. If the recipient has the correct key, they can decrypt the digest and verify the integrity of the data, and also the authenticity of the sender (via their shared key).

Signatures are the final piece of the puzzle – similar to HMACs, these use an algorithm to encrypt the digest; however, in this case, it is an asymmetric algorithm. The private key is used for encryption, and at the receiver end, the public key of the sender is used to decrypt and verify the integrity. Using robust principles of **public key infrastructure** (**PKI**), public keys can be trusted (their ownership can be verified).

The following table summarizes the differences between the three types:

Objective	Hash	HMAC	Signature
Integrity	✓	✓	✓
Authenticity		✓	✓
Non-repudiation			✓

Table 1.1 – A comparison of digest types

Transport security

The **TLS** protocol is a transport-level cryptographic protocol to ensure secure communications over a TCP/IP network. An encrypted transport layer is essential for APIs to ensure that attackers are unable to eavesdrop on data or tokens over the network and to ensure that the client can validate the identity of the server (via certificate validation). Certificate management has usually presented challenges to organizations; however, with the emergence of providers such as Let's Encrypt, certificate deployment and management have become a lot simpler.

Encoding

The final building block is that of **encoding**, which involves changing the representation of data for the purposes of storage or transmission. Encoding converts the character set of input data to a format that can be safely stored or transmitted, and decoding converts that data back to its original format.

This concept is best understood by looking at a few common encoding schemes:

- **HTML encoding**: In HTML, certain characters have special significance – for example, < and >. If a text block contains these characters, it will change the structure of the rendered HTML, which is undesirable. By encoding these special characters in another format (`"&lt;"` and `"&gt;"`), they can be safely rendered in an HTML document, where they will be displayed correctly as < and > but stored in a different form.

- **URL encoding**: Similarly, in a URL, only the ASCII character set is allowed; all other characters are forbidden. Unfortunately, path locations may contain such characters (spaces and underscores, for example). By encoding these to an ASCII text representation, it is possible to get a valid URL version – for example, a space is converted to %20.

- **ASCII, UTF8, and Unicode**: Text can be represented in a number of different formats and can be converted from one to the other, depending on the platforms and locales in use.

- **Base64**: This is a commonly used encoder to transform binary data to text data suitable for transmission over HTTP.

Encoding does not use a key to perform the transformation but, rather, a fixed algorithm, and any content that has been encoded can be decoded to produce exactly the same original content.

> **Encoding versus encryption**
>
> A common misunderstanding is the difference between encoding and encryption. They are two very different topics, solving different problems.
>
> Encoding transforms data from one representation to another using a fixed algorithm. No keys are used, and the encoded data can be trivially converted back to the original format. It does not offer any form of integrity or confidentiality functions.
>
> Encryption performs a transformation of data using a key; the resultant output does not resemble the input at all, and the only way the original data can be obtained is by applying the reverse decryption function using the same key. Encryption does not transform the representation or character set of the data.

Examining API data formats

Finally, in this section, let's take a quick look at common data formats used in APIs. For REST APIs, information is transferred in plain text format (although this information may be encoded), either as key-value pairs as request parameters, one or more headers, or as an optional request body. Responses consist of a status and an optional response body.

XML

eXtensible Markup Language (**XML**) is the original heavyweight format for internet data storage and transmission. The format is designed to be agnostic of data type, separates data from presentation, and is of course extensible, not being reliant on any strict schema definition (unlike HTML, which uses fixed tags and keywords).

Although XML was dominant several years ago, it suffered from some significant drawbacks, namely complexity and large data payloads. These two factors make it difficult to process and parse XML on resource-limited systems. XML is still encountered, although much less so in APIs.

A simple example of XML shows the basic structure of tags and values:

```
<note>
  <to>Colin</to>
  <priority>High</priority>
  <heading>Reminder</heading>
  <body>Learn about API security</body>
</note>
```

JSON

Javascript Object Notation (**JSON**) is now the dominant transfer format for data over HTTP, particularly in REST APIs. JSON originated as a lightweight alternative to the more heavyweight XML format, being particularly efficient with transmission bandwidth and client-side processing.

Data is represented by key-value pairs, with integer, null, Boolean, and string data types supported. Keys are delimited with quotes, as are strings. Records can be nested, and array data is supported. Comments are not permitted in JSON data.

A simple example of JSON shows the key-value pair structure:

```
{
  "name": "Colin",
  "age": 52,
  "car": null
}
```

YAML

YAML Ain't Markup Language (**YAML**) is another common internet format, similar to JSON in its design goals. YAML is in fact a superset of JSON, with the addition of some processing features. JSON can be easily converted to YAML, and often, they are used interchangeably, depending on personal preference, particularly for OpenAPI definitions.

The same data from the JSON example can be expressed in YAML as follows:

```
---
name: Colin
age: 52
car:
```

OpenAPI Specification

The final format we need to understand is the **OpenAPI Specification** (**OAS**), which is a human-readable (and machine-readable) specification for defining the behavior of an API. The OpenAPI Specification is an open standard run under the auspices of the OpenAPI Initiative. Previously, the standard was known as Swagger (aka version 2) but has now been formalized into an open standard, and currently, version 3.0 is in general use, with version 3.1 due imminently at the time of writing.

An OAS definition can be expressed either as YAML or JSON and comprises several sections, as shown here:

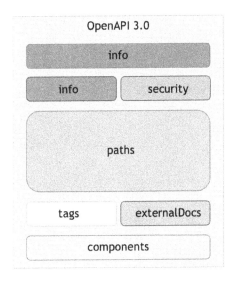

Figure 1.1 – OpenAPI Specification sections

Using an OAS definition at the inception of the API life cycle (referred to as *design-first*) offers several key benefits, namely the following:

- **Description validation and linting**: Parsers and audit tools can automatically validate a definition to confirm its correctness and completeness.

- **Data validation**: Request and response data can be fully specified, allowing validation of API behavior at runtime.

- **Documentation generation**: Documentation can be automatically generated from a definition, including a test UI, allowing the API to be exercised.

- **Code generation**: Tools exist that allow the server and client code stubs to be generated in a variety of languages, easing the burden on developers.

- **Graphical editors**: Fully featured graphical editors make it a simple task to design OAS specifications in an interactive, intuitive manner.

- **Mock servers**: OAS definitions can be used to build mock servers that simulate the behavior of an actual API backend. This is extremely useful in the early stages of API development and integration.

- **Security analysis**: Most importantly for us is the security benefits that the use of an OAS definition brings – definitions can be examined for security constraints (authorization and authentication, for example), and deficiencies can be highlighted. Data structures can be fully specified to allow the validation of data, preventing excessive information exposure.

A sample OAS definition is shown in the following snippet. This is an example of a bare-minimum specification of an API and includes the following in the header section:

- The OpenAPI version

- Information metadata

- Server information, including the host URL:

```
{
    "openapi": "3.0.0",
    "info": {
        "version": "1.0.0",
        "title": "Swagger Petstore",
        "license": {
            "name": "MIT" }
    },
    "servers": [
    {
      "url": http://petstore.swagger.io/v1
    }],
    ..
```

The next section in the OAS definition describes an endpoint, showing details such as the following:

- The endpoint path name

- The HTTP method to be used

- Request parameters

- Status codes

- The response format:

```
    "paths": {
        "/pets": {
            "get": {
                "summary": "List all pets",
```

```
                    "operationId": "listPets",
                    "parameters": [
                 { "name": "limit",
                   "in": "query",
                   "description": "Maximum items (max 100)",
                   "required": false,
                   "schema": {
                     "type": "integer",
                     "format": "int32"
                   } } ],
                  "responses": {
                    "200": {
                      "description": "A paged array of pets",
                      "headers": {
                        "x-next": {
                          "description": "Next page",
                          "schema": {
                            "type": "string
                        } } },
                      "content": {
                      "application/json": {
                        "schema": {
                        "$ref": "#/components/schemas/Pets"
                      } } } },
                      . .
```

At this point, we understand the building blocks of APIs and the associated data formats. It is now time to look at the elements of API security.

Understanding the elements of API security

API security is a complex topic and comprises many elements — a successful API security initiative should be built upon a solid foundation of a DevOps practice and a balanced AppSec program. Just like a house, the strength of the overall structure is dependent on a solid foundation – without these in place, an API security initiative may prove challenging.

Good security is built on a multi-layer system – this is the *defense in-depth* approach.

It is important to remember that API security is quite different from what has come before with web application security. This means that using existing tools and practices may be insufficient to produce secure APIs. Dedicated API security solutions must be deployed in addition to traditional AppSec tools to provide the optimum coverage and protection specific to APIs.

The elements of the API security hierarchy are shown here:

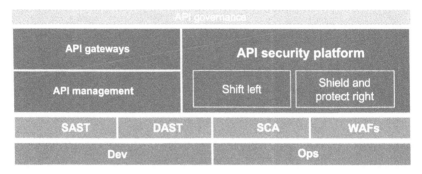

Figure 1.2: The elements of API security

Let's explore each of the layers of API security briefly.

DevOps

DevOps is a well-established set of practices to facilitate modern software systems, characterized by close relationships between the development and operations teams to improve methodology and practices and leverage the benefits of automation. DevOps is considered a continuous process with continuous improvements across several key domains in the **Software Development Lifecycle (SDLC)**, as shown here:

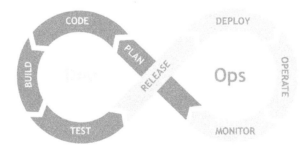

Figure 1.3: The DevOps cycle

DevOps offers many benefits to the delivery of software, including the following:

- Improved collaboration and trust
- Faster release cycles
- Reduced time to repair
- Higher levels of automation
- Use of standard processes, including testing and deployment

From the perspective of API security, the key benefit of DevOps is the ability to build APIs in a deterministic fashion using a standard process. Using standard **Continuous Integration / Continuous Delivery (CI/CD)** pipelines, API security testing and validation tooling can be injected into the build process to ensure that all deployed APIs have had the specified security checks and controls applied to them. APIs by their nature are well suited to automated testing, and the CI/CD pipeline is the ideal place for this activity.

SAST, DAST, SCA, and WAFs

Static application security testing (SAST), **dynamic application security testing (DAST)**, **software composition analysis (SCA)**, and **web application firewalls (WAFs)** form the vanguard of traditional application security programs.

The security of any software can be improved by the judicious use of such tools, as follows:

- SAST can detect basic flaws in source code at the time of development
- DAST can detect application vulnerabilities at runtime
- SCA can detect the use of vulnerable components and libraries
- WAFs can afford some level of protection against certain attack types

SAST can detect common coding vulnerabilities in API code (such as injection flaws) but will not detect API-specific flaws (such as broken authentication or authorization), since the SAST engine does not have contextual awareness of the underlying API code. Similarly, DAST is able to detect certain API vulnerabilities (such as a lack of rate limiting) but lacks the context to understand the API requests and responses.

WAFs are a mature technology for protecting web applications and offer some protection for APIs as well. They operate in line with traffic utilizing a so-called *allow list* to block suspected malicious traffic and allowing everything else. They can be configured to operate in monitor mode (passive) or blocking mode (active).

Organizations typically have dedicated security teams tasked with deploying and operating these tools within development teams. These teams should evaluate dedicated API security tools to complement some of the gaps that exist with these tools.

API management and gateways

API gateways are the workhorse of the API industry, providing a unified external interface to public clients and traffic routing to the relevant internal API backends after having performed transformation and conversion. Gateways are also responsible for network-level controls such as SSL termination, rate-limiting, IP address restrictions, and load balancing. Gateways can also implement security features such as JWT validation and identity management.

Some of the shortcomings of API gateways include the following:

- API gateways provide a central point of entry for API traffic and are effective at acting as a gatekeeper at the *front door* of the customer infrastructure; however, they are less effective at protecting what goes on behind the door

- Gateways are ineffective at protecting against several of the OWASP API Security Top 10 vulnerabilities

- Gateways can be inefficient at providing security processing functions such as traffic inspection

Typically, API management portals provide a level of API management on top of a gateway, allowing organizations to control their inventory, versioning, life cycle, and end-user experience by providing API catalogs.

Some of the shortcomings of API management platforms include the following:

- APIM portals are effective for providing a central view of an API inventory and also a single point of deployment for API policy

- Effective APIM deployment is contingent on development teams embracing a design-first approach and enrolling their APIs into a central portal

Both API management portals and gateways are vital components of an API security strategy, but their limitations should be borne in mind as part of the overall strategy.

API security platforms

The growth of API adoption has spawned several dedicated API security platforms, with the specific intent of addressing API security as a first-class citizen.

These platforms take different perspectives of securing APIs, including the following:

- Continuous monitoring of API traffic to detect emergent threats using **machine learning** (**ML**) and **artificial intelligence** (**AI**) technology

- Dedicated API firewalls that can protect APIs by enforcing the OpenAPI contract – this is the positive security model covered in the next section

- Scanning APIs to validate the API behavior against an OpenAPI contract

- Providing audit tools to ensure OpenAPI contracts adhere to best practices for data and security

Dedicated API security tools are vital to providing the final layer of API security. Now that we understand the elements of API security, let us conclude this chapter by setting API security goals.

Setting API security goals

Finally, in this chapter, let's focus on the security goals that should be considered in API security initiatives. Different organizations will have different security priorities based on their business priorities – a financial service organization will favor high levels of security and strict governance, while a social media portal may have lower security requirements and favor feature delivery instead. No two organizations have the same goals.

The three pillars of security

The term API security has a broad scope, meaning different things to the beholder. IT security has traditionally used the *CIA triad* to characterize risks to systems. **CIA** is an acronym for **Confidentiality, Integrity, and Availability**, and has applications in APIs as follows:

- **Confidentiality**: For APIs, this implies that data is transmitted using secure transmission channels (typically, TLS) and that only permitted clients are able to access resources belonging to them (enforced by access controls).

- **Integrity**: For APIs, this requirement ensures that data cannot be modified or tampered with by unauthorized parties. Again, TLS and access controls are critical to ensuring integrity.

- **Availability**: APIs should be resilient and resistant to DoS attacks designed to take an API offline.

While a useful framework for considering API security, it should be considered in combination with the OWASP API Security Top 10 covered in *Chapter 3*.

Abuse and misuse cases

When considering API security, we primarily consider hacks or breaches where an adversary deliberately attacks an API and causes it to misoperate, due to inherent flaws. Such attacks are deliberately focused on using techniques we will explore in the *Attacking APIs* section.

However, there is another category of API security risk to be considered – namely, the abuse and misuse of APIs. Typically, in this category, we consider automated scripting, bot attacks, scrapers, and nuisance actors. While they do not have a high-risk rating (according to the CIA triad, for instance), they can have detrimental consequences for organizations.

Some typical examples include the following:

- Bots attempting to enumerate APIs and discover endpoints

- Scrapers trying to exfiltrate large volumes of data through automated pagination (typically, online retailers or estate agents are targets)

- Spammers or so-called *troll farms* abusing social media APIs

- Nuisance actors being mischievous by using APIs in unusual or unexpected ways (such as the automation of online auction sites)

Some of these types of abuse cases can be relatively difficult to either defend against (because they appear to be no different from normal users) or to detect.

Data governance

Data governance is tangential to API security but a key consideration for a holistic API security strategy. APIs are primarily conduits for data transfer between internal systems or organizations and consumers or partners. APIs simplify the ability of developers to expose increasing amounts of data almost at the click of a button. However, with this ease comes an increased risk of inadvertent or unintended data leakage, causing regulatory and compliance concerns.

A solid data governance program is essential to ensure that consumers (typically, API developers in this context) have full awareness of the data sensitivity and classification and apply the relevant controls to limit access, in line with regulatory and compliance concerns.

This is particularly applicable to the financial services and the medical industry, which increasingly face data disclosures via APIs.

A positive security model

Unlike web or mobile applications, APIs present a tremendous opportunity to radically shift the security paradigm. Traditionally, web or mobile security has relied on a *negative security model*, which means that a known bad actor is blocked, and everything else is allowed. Here, a *deny list* approach is used.

This approach – while long-established – has a significant disadvantage in that defenders do not know the full extent of all known bad actors. Clever attackers can construct payloads or inputs that appear to be valid inputs passing through the deny list; however, in the context of the application, they are dangerous. Think of the example of **SQL Injection (SQLi)** attacks where seemingly innocuous input is applied to a database, where it can have catastrophic consequences. The negative security model is characterized by both high false positives and false negatives.

API security turns this model around entirely, relying instead on an allow list that passes only known good actors to the API backend. This is the *positive security model*, which only allows data and operations specified by the OpenAPI contract to access the API. Anything else is simply blocked (via an API firewall, for example) before reaching the API. This approach offers a massive benefit for security — the instances of both false positives and negatives are greatly reduced. The positive security model has one major drawback, however – it is reliant on a fully formed OpenAPI contract to operate correctly. This may be challenging for organizations not embracing an API design-first approach. However, the positive security model promises to be game-changing for the world of API security.

Risk-based methodology

Finally, let's conclude with an approach for prioritizing API security initiatives, which can be costly and time-consuming in large organizations. Probably the most frequently asked question is "*Where do I start?*" – security leaders are often stuck in a quandary when faced with a choice of trying to address their entire API portfolio (at great cost and a higher likelihood of failure), erroneously focusing on less important APIs (and wasting valuable security resources), or in extreme cases, simply not starting at all due to the enormity of the undertaking.

A common-sense approach to prioritizing an API security initiative is to use a risk-based methodology – start with the highest-risk APIs and work through to the lower risks, as budget permits.

Priority is dependent on several parameters, typically the following:

- **Network access**: Is the API publicly exposed, or is it on a more restricted network?
- **Data sensitivity**: How sensitive is the data and, hence, the impact of leakage? **Personally Identifiable Information** (**PII**) data (typically medical and financial data) is the highest sensitivity.
- **Access control**: Finally, how well protected is the API via access controls? Unauthenticated APIs are obviously the highest risk and should only be used for publicly accessible data.

Combining these three factors allows us to gain an approximate risk-based priority:

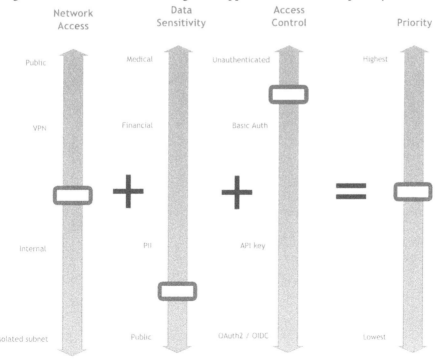

Figure 1.4 – Prioritizing API security via a risk profile

As a (slightly contrived) example, an unauthenticated API on a public network conveying medical records scores the maximum risk and, hence, becomes a priority.

While this is, at best, an approximate risk rating, it serves well to focus security activities where they will achieve the maximum return on investment.

Summary

We have covered a lot in this first chapter. APIs are the lifeblood of a modern application economy and, unfortunately, are a favorite target for attackers due to the high value of data that they convey. We now understand the core building blocks of APIs, how developers are vital in the journey to building secure APIs, and the various elements of a secure API initiative.

In the next chapter, we are going to explore in greater detail exactly how an API is constructed, how they work, and critically, how they are secured.

Further reading

- For further perspectives on the API economy, the following provide good insight:

  - https://marker.medium.com/the-api-economy-in-finance-payoffs-of-getting-connected-c5c6aeb34c57

  - https://www.forbes.com/sites/forbestechcouncil/2021/09/10/how-to-succeed-in-the-api-economy/?sh=a2b84a544b98

  - https://nordicapis.com/20-impressive-api-economy-statistics/

- The following will help to understand why SAST and DAST tools are insufficient to secure APIs:

  - https://thenewstack.io/application-security-tools-are-not-up-to-the-job-of-api-security/

  - https://restfulapi.net/

- The following will help to understand how Airbnb implements its data governance:

  - https://medium.com/airbnb-engineering/data-quality-at-airbnb-e582465f3ef7

- Further information on the different types of APIs:

  - `https://developer.mozilla.org/en-US/docs/Web/HTTP`

  - `https://www.altexsoft.com/blog/soap-vs-rest-vs-graphql-vs-rpc/`

- OpenAPI specification documentation:

  - `https://spec.openapis.org/oas/v3.1.0`

  - `https://spec.openapis.org/oas/v3.0.0`

2

Understanding APIs

With our understanding of the role APIs play and the security challenges associated with them, let's take a deeper look at how APIs work in much more detail. By the conclusion of this chapter, you will have a full understanding of how APIs work, starting with the transport protocol and the choice of APIs available to developers. The two central tenets of API security are *authentication* and *authorization*— we will cover these in detail to understand their pivotal role. Finally, you'll understand how keys and tokens are used in ensuring the security and integrity of APIs and the identity of API clients.

In this chapter, we're going to cover the following main topics:

- The fundamentals of HTTP
- The different types of APIs that are currently in use
- Authentication best practices and methods
- Authorization best practices and methods
- Using tokens and keys

Understanding HTTP fundamentals

APIs are built on top of the **Hypertext Transfer Protocol** (**HTTP**), which in turn uses the **Transport Control Protocol** (**TCP**) as a transport layer providing guaranteed error-free data delivery. HTTP was originally designed for the transfer of hypertext documents (such as HTML files) but has been adapted for many other uses due to its ubiquity across systems and because it is generally accessible through firewalls and routers, avoiding the use of custom ports or protocols.

HTTP comprises a few core elements, which we will discuss in the following sections.

Uniform Resource Locator

The **Uniform Resource Locator** (**URL**) is the *address* of a resource (file, JSON record, image, etc.) on the internet. URLs are unique (can only reference a single resource) and are fully qualified (meaning they can be resolved to the resource location without ambiguity).

The generic form of a URL is shown here:

```
scheme://host[:port]/path[?query-string][#fragment-id]
```

The elements are as follows:

- **Scheme**: The scheme specifies the protocol to be used to access the resource. Schemes comprise a keyword and a *://* separator. Common protocols include `http`, `https`, `ftp`, and `file`.

- **Host**: The host specifies the location of the host containing the resource, typically a server. The host must be resolvable to a network location via the client, typically using DNS. Hosts can be local or **fully qualified domain names** (**FQDNs**) on the public internet and may also use a combination of local names plus FQDNs, for example, `server1.mydomain.com`.

- **Port**: The optional port parameter specifies the **Internet Protocol** (**IP**) port to be used on the host. If not specified, the default ports will be used—port `80` for `http` and port `443` for `https`. Using ports allows multiple servers to be hosted on the same physical network device.

- **Path**: The path provides the location of the resource on the host storage system. This is directly analogous to a file path on a local disk drive: paths are hierarchical and nested.

- **Query string**: The optional query string allows the request to be customized to provide parameters to the server to refine the resource(s) to be returned. Queries are preceded by the ? operator and separated by the & operator; key and value pairs are separated by the = operator. The following is an example of a query string:

  ```
  ?first_name=Homer&last_name=Simpson
  ```

- **Fragment ID**: Finally, the optional fragment ID specifies a location within the resource; typically, this is used to locate section headings within an HTML document and is rarely used for APIs.

The elements of a URL create a reference to a resource on the internet. Next, let us understand how to perform data transfers to and from the resource.

Requests

HTTP uses plain text for data transfer between the client and server; the following elements are included in a **request**:

- A *request line* containing the HTTP method, the relative path, and the version of the HTTP protocol.

- The *header section* containing colon-separated key-value pairs, each one on a new line. These headers control various request parameters and protocol behaviors. A blank line is inserted after the last header.

- The *body* of the request in plain text, such as JSON, XML, or YAML, or a text-encoded version of binary content.

An example HTTP POST request to a server is shown here:

POST /api/login HTTP/1.1
Content-Type: application/x-www-form-urlencoded
User-Agent: PostmanRuntime/7.29.0
Accept: */*
Postman-Token: 43c514cb-a765-41ea-897a-0f74bfe6ff6c
Host: localhost:8090
Accept-Encoding: gzip, deflate, br
Content-Length: 35
Connection: keep-alive

pass=hellopixi&user=user%40acme.com

Figure 2.1 – Sample HTTP request

In the preceding request, the client is making a request to a server to perform a login by providing a username and a password. The key elements of this request are as follows:

1. The HTTP method—in this case, it is a POST request, which indicates that data will be included in the request body.

2. The relative server path is /api/login—note the hostname is not required in this field since this is used at the transport level in the TCP request.

3. The HTTP protocol version to be used—HTTP 1.1 in most cases.

4. A number of HTTP headers, including Content-Type (telling the server the nature of the body content), Host (the originating hostname), and Content-Length (the length of the request body).

5. The request body, which in this case is a *URL-encoded* password and username key pair (note the @ is converted into its corresponding hexadecimal value by the URL encoding).

Once the body has been transferred, the client will await a response from the server.

Responses

The HTTP **response** is again a plain text payload with the following elements included in the response:

- A *response line* containing the version of the HTTP protocol, and an HTTP response code pair (numeric code and a string representation).

- A *header section* containing colon-separated key-value pairs, each one on a new line. These headers specify various response parameters and protocol behaviors.

- The *body* of the request in plain text, such as JSON, XML, or YAML, or a text-encoded version of binary content.

An HTTP response to the sample request (see *Figure 2.1*) from a server is shown here:

```
HTTP/1.1 200 OK
X-Powered-By: Express
Content-Type: application/json; charset=utf-8
Content-Length: 859
ETag: W/"35b-GGj5vhMkH+Jh4NW8P7JMgTTitLM"
Date: Sat, 09 Jul 2022 14:23:25 GMT
Keep-Alive: timeout=5
Connection: keep-alive

{"message":"Token is a header JWT
","token":"eyJhbGciOiJSUzM4NClsInR5cCl6lkpXVCJ9
```

Figure 2.2 – Sample HTTP response

The key elements of this response are as follows:

1. The HTTP protocol version used—HTTP 1.1 in most cases.

2. The HTTP response code pair (numeric code and a string representation). In this case, a 200 OK was received indicating the request was processed successfully.

3. A number of HTTP headers including Content-Type (telling the client the nature of the body content) and Content-Length (the length of the response body).

4. The response body, which in this case is a JSON payload with a message field and a token field.

The client will verify that the response body matches the Content-Length field and then begin processing the response.

Methods

As mentioned in the preceding sections, HTTP requires a client to specify the HTTP **method** (sometimes called the *operation*) to be used in the request.

A key concept in selecting the correct method is to understand the impacts of making multiple requests that are the same. Internet transport can be intermittent, and clients may attempt to repeat an operation if no response is received within a timeout window. If the server receives a duplicated request, it is important to understand the server's behavior. This is the concept of **idempotency**—an operation that is considered *idempotent* can be repeated multiple times without changing the state of the resource on the server. In contrast, a *non-idempotent* operation will make changes to server resources each time it is invoked, even if it is the same request content.

This concept is best understood using examples: a GET request is *idempotent*—no matter how many times the GET request is invoked, the resource on the server is not changed. A POST method, however, is *non-idempotent*—each call to a POST method will create a new resource on the server, even if the request is the same.

Method	Description	Idempotency
GET	GET is used to request the specified resource from the server.	Idempotent
POST	POST is used to send data to a server to create or update a resource. A new resource will be created on each call.	Non-idempotent
PUT	PUT is used to send data to a server to create or update a resource. Multiple requests will not create new resources.	Idempotent
HEAD	HEAD is almost identical to GET, but the server does not return the response body. This can be used to test whether the resource exists before calling GET to retrieve it.	Idempotent
DELETE	The DELETE method deletes the specified resource if it exists; it does nothing otherwise.	Idempotent
PATCH	The PATCH method is used to make partial modifications to a resource. This method is not supported on all servers.	Non-idempotent
OPTIONS	The OPTIONS method allows the client a way to identify all the request methods supported by the server in advance of invoking them.	Idempotent
CONNECT	The CONNECT method is used to start a two-way tunnel with the requested resource.	Non-idempotent
TRACE	This is a debugging method that is used to perform a message loop-back test that echoes back the client's request.	Idempotent

Table 2.1 – HTTP methods

For APIs, the most used methods are the POST, GET, PUT, PATCH, and DELETE methods since these provide the primitives necessary to implement the **Create, Read, Update,** and **Delete (CRUD)** operations necessary for common data operations. To create a resource, the POST or PUT methods are used; to read a resource, the GET method is used; to update a resource, the PUT or PATCH methods are used; and to delete a resource, the DELETE method is used.

For developers, the POST operation requires caution since this may have the inadvertent side effect of creating duplicate resources.

Status codes

Each HTTP request returns a single **status code**, both in integer format and a descriptive text message. Clients should check the status code to determine whether the operation succeeded as expected; whether a fatal error occurred that cannot be recovered and indicates permanent failure; or whether an error occurred that may be recovered with a retry of the request.

The top-level error codes are shown in summary here:

Code ranges	Type of response	Details
100–199	Informational responses	Infrequently used, these codes indicate an interim result to the client.
200–299	Successful responses	Indicates a successful request. 200 is OK and used by all methods, and 201 is Created and used by the POST method.
300–399	Redirection messages	Indicates to the client that the requested resource has moved to another location and that a subsequent operation is needed. 301 is moved permanently and is the most encountered.
400–499	Client error responses	Indicates that the client had an error in the request. 400 is Bad request, 401 is Unauthorized, 403 is Forbidden, and 404 is Not Found. Clients should handle individual cases separately.
500–599	Server error responses	Indicates that the server encountered an error processing the request. 500 is Internal Server Error, 501 is Not Implemented, and 502 is Bad Gateway. Server errors may be transient and might succeed on a retry.

Table 2.2 – HTTP status codes

Sessions

The final topic to explore in relation to HTTP is that of **sessions** or the *state* of a client/server interaction.

Each HTTP request is self-contained and does not require knowledge of preceding or subsequent requests — for a given request, the server has no knowledge as to whether that client has made previous requests. This is a *stateless* transaction.

For a rich client experience, it is important that the state be maintained between sequential operations to allow clients to implement state management, that is, navigation history, contents of shopping carts, search results, and so on. Web sessions can be made *stateful* using **cookies**, which are small text files transferred from the server to the client on the first request and stored and presented on subsequent requests to the domain. By using cookies, servers can track clients via the contents of the cookie and render the content relevant to that session state—this is a *stateful* session. Cookies, while powerful, present several security risks since, once an attacker has stolen a cookie, they can impersonate the associated client.

Another approach to managing the state is to include a token in the HTTP request headers that can be used by the server to track the state of the client—by including this token in the request, it is possible to secure the entire transaction using transport security.

APIs are generally considered to be *stateless*, with each operation being decoupled from the preceding operations.

Exploring the types of APIs

Given our understanding of the basics of HTTP, let's explore the most common types of API currently in widespread use. We will take a deeper look at the basics of the protocols, their typical use cases, and their strengths and weaknesses. Let's dive in!

REST

REST is the acronym for **REpresentational State Transfer** and an architectural style for transferring hypermedia content over HTTP. REST does not specify a rigid standard, but rather a guideline to be followed, encapsulated by six principles:

- **Uniform interface**: The server should present a uniform interface to the client – namely, the interface should be identical across all client devices and platforms.

- **Client-server**: There are two clear roles in REST: a client requesting and receiving data, and a server responding and sending data. No other roles are defined.

- **Stateless**: REST is stateless, meaning that each individual request does not depend on preceding or subsequent requests. The onus is on the client to present all state context in each individual request and not be reliant on the server to manage the state.

- **Cacheable**: The server is responsible for labeling data as either cacheable or non-cacheable. The client should honor such declarations and not make assumptions.

- **Layered system**: Complexity is managed by using a hierarchical layered architecture where layers offer discrete functionality, which can be composed into higher-order operations.

- **Code on demand** (optional): Client functionality can be extended by executing code on the client. This is not relevant to APIs.

REST is by far the widest used of the API types, accounting for approximately 75% of API traffic in 2022. The overwhelming reason for this popularity is the *simplicity* of REST APIs—data is transferred in common, easy-to-use formats in plain text, the design maps to the underlying REST architecture (the Create-Read-Update-Delete model), and the client-server model allows for a high level of abstraction and decoupling.

Having said that, there are some significant disadvantages that should be considered regarding REST as a choice for APIs:

- **No single REST structure**: The simplicity of REST means that few constraints are placed on API designers, who are free to choose their own data models and interface designs. Each REST API is unique, and this can lead to difficulty in adoption unless they are well documented.

- **Large payloads**: While the use of rich plain text payloads contributes to design simplicity, it comes at the expense of relatively large payloads, particularly for more complex or richer queries. This may impact performance on slower mobile networks.

- **Over- and under-fetching of data**: A related issue with REST APIs is the fact that the client is responsible for fully specifying the parameters of the query to retrieve the required data. Poor data or API design can result in clients that either fetch too much or too little data and need to perform multiple operations to retrieve the desired data.

To understand how a REST API is constructed, consider the following example:

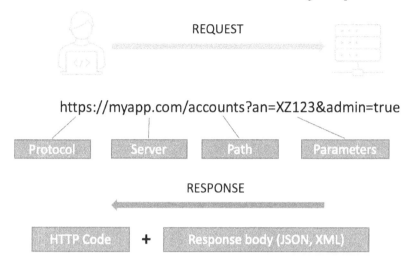

Figure 2.3 – Example REST API request

In this example, the client constructs a URL specifying the protocol, the server, the API endpoint (in this case, the *accounts* resource is being queried), and parameters to specify the *account* to be retrieved, (in this case, an account number of YZ123). The response returns an HTTP status code and a response body in JSON, XML, or YAML.

GraphQL

GraphQL is a newcomer to the API landscape and was developed by the engineering team at Facebook to better serve graph data (as is typical on a social media platform). GraphQL attempts to address one of the key issues with REST APIs: over- and under-fetching data. With REST APIs, the client must construct the queries to build a nested data view (for example, a list of friends, their interests, a home city, etc.) by making a sequence of calls to a variety of different endpoints. This can result in excessive querying (over-fetching) or missing data (under-fetching).

GraphQL addresses this problem by shifting the burden of data processing to the server, allowing complex data structures to be retrieved in a single API call. A client specifies the data model (using the GraphQL language), and the server populates the response accordingly.

The key advantages of GraphQL include the following:

- **Precise queries**: Queries can be constructed to return exactly the data that the client requires – no more and no less.

- **Nesting**: Complex nested data records can be retrieved since GraphQL is graph-focused rather than resource-focused like REST.

- **Strongly typed**: Data types are strongly specified in the query so the client gets exactly the data it expects.

- **Discovery**: GraphQL endpoints can be dynamically queried to allow users to discover the data models underpinned by the endpoint.

GraphQL does suffer from a few significant drawbacks, many of them relating to security. Firstly, servers are complex and allow dynamic discovery, which opens up a greater attack surface and the possibility of denial-of-service attacks. Secondly, servers may over-expose data and make it difficult to apply fine-grained access control models.

To understand the value of GraphQL, let's take a sample query to return some details from the `"react"` repository in the `"facebook"` organization on GitHub, as shown:

```
query GetRepositoryWithIssues {
  repository(owner: "facebook", name: "react"){
    id
    nameWithOwner
    description
    url
    createdAt
    issues(last: 2) { totalCount
      nodes{
        title
        createdAt
        author {
          login
        }
}}}}
```

In summary, the query asks for some basic repository information and then the first two issues, including the author's name. If this query was constructed in a REST API, it would require several calls to the `repository`, `issues`, and `author` resources.

The response data is shown here:

```
{
  "data": {
    "repository": {
```

```
      "id": "MDEwOlJlcG9zaXRvcnkxMDI3MDI1MA==",
      "nameWithOwner": "facebook/react",
      "description": "A declarative, efficient, and flexible
JavaScript library for building user interfaces.",
      "url": "https://github.com/facebook/react",
      "createdAt": "2013-05-24T16:15:54Z",
      "issues": {
        "totalCount": 11552,
        "nodes": [
        {
          "title": "Bug: React 18
renderToPipeableStream              missing support for nonce for
bootstrapScripts and bootstrapModules",
          "createdAt": "2022-07-09T21:01:51Z",
          "author": {
            "login": "therealgilles"
          }
      } } } } }
```

As shown, we can see that the GitHub server has populated the query with the appropriate response data, limiting the number of issues to 2 as specified. A single query can retrieve a relatively complex data structure.

RPC

A **Remote Procedure Call** (**RPC**) API differs from others in that it allows for the remote execution of code on the server—the client prepares a procedure call using a stub library, which is packaged into a universal format for transmission on the network. On the server, the payload is unpacked and then invoked on the server stub, and finally, the response is packaged and returned to the client.

RPC APIs are predominantly used for command-oriented applications where actions need to be executed remotely—it is more an action-oriented protocol than, say, the REST of GraphQL, which is data-oriented. The Google implementation **gRPC** is the most common variant in use currently.

SOAP

The **Simple Object Access Protocol** (**SOAP**) is the oldest of the API types in common use, dating back over two decades. SOAP is a service-oriented, XML-formatted, highly standardized web communication protocol. SOAP messages contain up to four components, including an envelope, a header, a body, and faults (errors). SOAP is also highly extensible and allows multiple extensions, and includes WSDL for client library design.

SOAP suffers from high complexity in implementation and has the largest data payloads of all API types.

WebSockets

The **WebSockets** API is built on top of the WebSocket communication protocol, which provides full-duplex transmission over a TCP connection. The primary advantage of the WebSocket protocol is a lower latency of communication than a corresponding half-duplex REST API connection. This API/protocol is popular within modern browsers to provide media streaming services.

> **Important note**
>
> In this book, unless otherwise stated, we will be referring to REST APIs. The exception will be dedicated sections focused on GraphQL. RPC and SOAP are not covered further in this book at all.

Access control

The foundation of API security is access control, which comprises two elements—authentication verifies who accesses an API, and authorization determines what they may do.

> **Beware of mixing up authentication and authorization concepts**
>
> Although the two concepts are closely related and used in conjunction to provide secure and granular access to an API, they are distinctly different. *Authentication* is about establishing the identity of a user or client and whether they are who they say they are, usually by means of a set of credentials. *Authorization* determines what access to resources a user or client is permitted based on who they are (established via authentication).

No authentication

Public APIs intended for read-only access may not require authentication if they are intended for anonymous (unauthenticated) access. Typically, these endpoints are used for information or status, for example, a status API for an online service.

HTTP authentication

Dating back to the early days of the **World Wide Web** (**WWW**), *HTTP authentication* uses a sequence of HTTP calls to get access credentials for a restricted resource (such as a private web page) and then verify those credentials to allow (or deny) access.

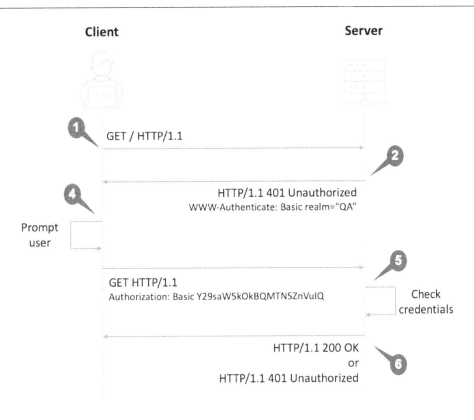

Figure 2.4 – HTTP authentication sequence of calls

The preceding exchange proceeds as follows:

1. The client browser makes a request to a protected resource.

2. The server responds with a 401 Unauthorized response and a WWW-Authenticate header with the security realm (specific to the server).

3. The client will then either prompt the user for a username and password combination or, if a stored token is present, it will make a GET request with an Authorization header with the token value.

The server will then validate the token against the valid credentials and either return 401 Unauthorized if the credentials were invalid or 200 OK if they were valid.

Let us look at two network traffic logs from the Chrome browser to understand the detailed workings of this protocol.

Firstly, a failed *login* attempt is shown here:

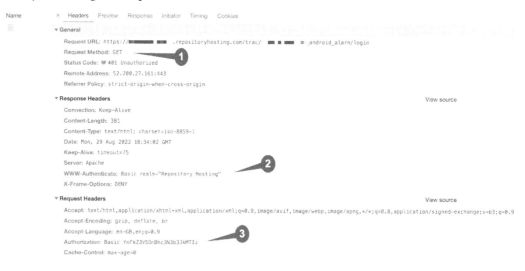

Figure 2.5 – Failed HTTP authentication

Here are the key things to note:

- The browser accesses the /login URL, makes a GET request, and receives 401 Unauthorized in return (this is step 1 in *Figure 2.4*).

- The server returns a WWW-Authenticate header specifying the realm – in this case, the company name (this is step 2 in *Figure 2.4*).

- The request from the browser also contains the Base64 representation of the username and password pair.

It is trivially possible to decode the Base64 value (YmFkZ3V5OnBhc3N3b3JkMTIz) to reveal the credential pair—in this case, *badguy:password123*.

Now let us examine a successful *login* attempt, as shown here:

Figure 2.6 – Successful HTTP authentication

Here are the key things to note:

1. The login request is precisely the same as that in *Figure 2.5*; however, in this case, a 302 Found response was returned. This indicates success and the server begins returning the home page of the application (note the web contents in the left panel).

2. Since the correct credentials were supplied, the server returns a fully populated response specific to the server type. In this example, the host is a Trac issue tracker server, which returns three cookies to manage authorization and tracking.

3. As per the failed login, the request from the browser also contains the Base64 representation of the username and password pair (in this case, they are obfuscated since they are valid).

There are two principal variants of this authentication: basic authentication and digest authentication.

Basic authentication

Basic authentication is largely outdated, originating in early website authentication. The client presents a *username* and *password* pair, which is encoded as a `Base64` string and transmitted in the request headers, and finally verified by the server.

The familiar popup for a basic authentication prompt dialog prompting for a username and password is shown here:

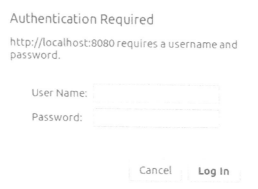

Figure 2.7 – Basic authentication pop-up dialog

There are several notable security issues with basic authentication:

- **Clear text transmission of credentials**: If the `Base64` username/password combination is transmitted over HTTP, it can easily be captured in transit.

- **Password stuffing attacks**: An attacker can easily brute-force the login using password dictionaries.

- **Vulnerable to replay attacks**: This is because the `Base64` username/password combination is identical for each computation on the same credential. By capturing this value, it can easily be replayed against a server at a later time.

- **Server spoofing**: The client has no way of validating the server's authenticity, allowing spoofing attacks.

- **Vulnerable to machine-in-the-middle attacks**: If HTTP is used, man-in-the-middle attacks are possible.

Given these weaknesses and the far more secure alternatives available, basic authentication is seldom used anymore.

Digest authentication

By using a hashed password value instead of transmitting the actual credential to the server, digest authentication overcomes many of the weaknesses of basic authentication. Hashing is a *one-way function* that cannot be reversed to retrieve the original input. The advantage of hashing is that the password is no longer transmitted in plaintext but rather a hash of the password. On the server, a hash is calculated using the same algorithm as the client against the correct password to produce a server hash. This is compared with the received hash—if they are identical, the client used the correct password and is thus authenticated. This eliminates the transmission of clear-text passwords.

A weakness of basic authentication is the vulnerability to replay attacks. This can be overcome by using a *nonce* (number used once), added to the password to be hashed. This results in a different hash for each request, meaning that even if the hashes are captured in transit, they will be invalid if replayed to a server.

AWS keyed-HMAC authentication

Amazon Web Services (**AWS**) has chosen to use a bespoke authentication mechanism based on **keyed-HMAC** over HTTP. The AWS signature authentication allows verification of the requester's identity, in-transit data protection, and the prevention of replay attacks.

Further information is available at `https://docs.aws.amazon.com/AmazonS3/latest/userguide/RESTAuthentication.html#ConstructingTheAuthenticationHeader`.

Session cookies

The traditional way for a browser client to maintain a state with a server is via the use of **cookies**, which are small JSON objects passed to a client by a server. On subsequent access to the same server, the client will present the cookies upon connection, allowing the server to identify the client's previous state. Intrepid web developers quickly learned that they could store authentication information as part of the session state, preventing the need for constant re-authentication by users.

As a natural evolution, when APIs started to dominate the backend landscape, it became expedient to reuse these **session cookies** to access the backend API. The most typical use of session cookies is in appliances or embedded devices where a web user interface is added on top of an appliance to allow management. Typically, this web UI will use session cookies to maintain the user's state, and then reuse these against the backend API.

The use of session cookies is not recommended as a best practice for API design. The first concern is the fact that session cookies can relatively easily be stolen by an attacker with access to the browser session using cross-origin script attacks targeting misconfigured servers. *Chapter 4, Investigating Recent Breaches*, contains an excellent example of a **cross-site request forgery** (**CSRF**) attack being used to gain access to an on-prem server cluster.

On top of this, session cookies present challenges in their lifetime management (how long should they last and how are they revoked?). Avoid using session cookies as far as possible.

API keys

Dedicated API keys were the natural evolution from the methods described previously, which were attempts to shoehorn web authentication methods into a new role in authenticating APIs. The workhorse of API authentication is a dedicated API key (also referred to as a token), which is presented to the API server on every request by the client, usually as part of the header. API keys are usually generated by a server or management portal and distributed to clients. A good example is the use of access keys for third-party SaaS services such as Twilio, GitHub, and cloud providers. The account owner is in control of the API key lifecycle and distribution.

API keys solve many of the challenges presented by the re-purposing of web authentication methods but they still present significant challenges to system designers: how are they to be stored securely on client devices, and how is their lifecycle to be managed? *Chapter 4, Investigating Recent Breaches*, presents several examples of breaches originating from the leakage of API keys. Best practice recommends the use of dedicated keys per client (rather than the reuse of a small number of keys) for the blast radius to be minimized upon breach. Always make sure the end user or owner has control of their API keys, and in particular, that they can easily revoke keys at their own will.

For embedded platforms or mobile devices, embedded API keys should be avoided due to the difficulty in re-issuing keys and the risk of leakage via device reverse engineering. API keys are suitable for APIs within a microservice architecture where their access can be controlled by platform policies and access controls.

OAuth 2.0

Having followed the natural progress of API authentication, we arrive at the current standard for modern web and API authentication: OAuth 2.0. OAuth had its genesis as OAuth1 at Twitter as far back as 2006, from where it evolved to become the standard for major cloud providers such as Twitter, Google, Facebook, and so on. It was notoriously difficult to implement an OAuth1 client.

Many providers began to deprecate OAuth1 in favor of OAuth 2, which has become the de facto standard for access control for API clients and the delegation of permission (authorization) to clients. In essence, OAuth 2.0 (henceforth referred to as simply OAuth) is a flow to allow a user to verify their identity to an authorization server (typically via a login page), which allows the requested access permissions (roles), and in turn, receive a token that is used in subsequent access. The client (for instance, a mobile device application) acts on behalf of the user, does not have direct access to the user's credentials, and only has specific access permissions afforded via a token. Authority is delegated in a limited scope, rather than allowing the client to have the same access as the account owner.

The easiest way to understand OAuth2 is to examine a well-known example: connecting one web service to another. In this example, I am going to connect my Dropbox account to my Gmail account.

Before looking at how this is implemented with OAuth2, let's consider how we would have linked two accounts together. Usually, this would have involved creating a login in Dropbox and then going to a page to connect applications and entering your credentials for the service being connected. There are several serious weaknesses with this approach:

- The most obvious concern is the fact that we are entering our credentials on another website, with no guarantee of them being securely stored or, worse, someone abusing them by accessing your account.

- Equally concerning is that we are delegating full access and privileges of our account to a third party. In our example, we probably only want Dropbox to access our contacts, but instead, we allow Dropbox full and unfettered access to our Gmail, allowing them to read or send emails!

- It is impossible to revoke access without changing your credentials. Imagine having to change your Gmail password every time you revoked an application's access.

OAuth2 solves all these problems and then some. Let's revisit this scenario using OAuth2.

Opening the Dropbox application presents a typical screen, as shown here:

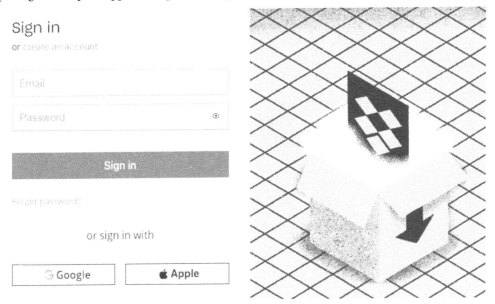

Figure 2.8 – DropBox login dialog

Since we have a Google account, we can use that to log in—in this case, Google is actually acting as the **identity provider** (more on this later).

Once we click the Google button, the Dropbox application redirects the user to Google via a pop-up web page with the dialog shown here:

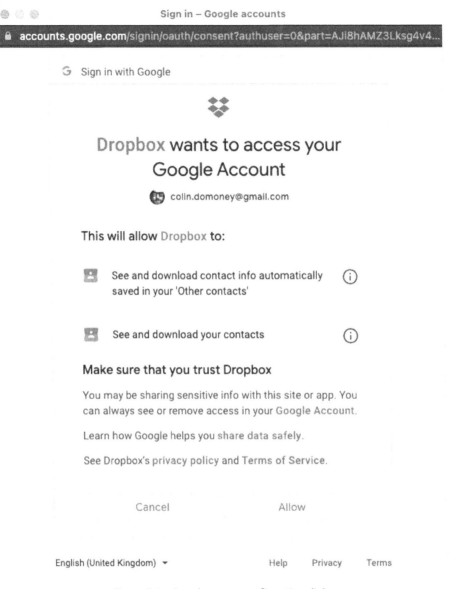

Figure 2.9 – Google access confirmation dialog

The key things to note in that dialog are Google explicitly tells you what rights are being given to Dropbox on your behalf, asks you to confirm that you trust Dropbox, and there are options for **Allow** or **Cancel**.

If you click **Allow**, Google will pop up a dialog confirming the activity and redirect to the Dropbox application, which is now connected.

Dropbox was granted access to your Google account

colin.domoney@gmail.com

If you did not grant access, you should check this activity and secure your account.

Check activity

You can also see security activity at
https://myaccount.google.com/notifications

Figure 2.10 – Google access confirmation dialog

Behind the scenes, Google has issued an access token to Dropbox, which effectively confirms our identity and the roles we have been assigned. Dropbox stores this token and uses it every time it wishes to access our Gmail account. That's how easy OAuth2 really is!

The previous weaknesses are solved as follows:

- We have not entered our Google credentials anywhere, least of all on a third-party site. This removes any chance of credential leakage or abuse.

- When Google prompted us to allow access, it explicitly listed the rights being assigned to Dropbox. If Dropbox attempted to read our email, this operation would be forbidden. As an end user, we are in control of the access we wish to delegate to a third party.

- Since Dropbox has been given an access token unique to this trust relationship between Google and Dropbox to read contacts, it is a simple matter of logging in to Google and revoking this access without impacting any other access, and it certainly does not require a global change of credentials.

Hopefully, by this point, you will appreciate the benefits of OAuth2. Let's now investigate the finer details of how the standard works. Firstly, we need to understand the various stakeholders in a scenario:

OAuth2 Role	Description	Example
Resource owner	The owner of the protected resource—usually the end user.	Gmail user
Resource server	The server hosting the protected resource.	Gmail application
Client	The application requiring access to the protected resource.	Dropbox
Authorization server	The server authenticating the resource owner and issuing access tokens if authorized by the owner.	Google ID server

Table 2.3 – OAuth2 roles

To ensure the flexibility and extensibility of the OAuth framework, several different flows (grant types) are provided to claim access tokens from the server. The choice of flow depends on the client type. Certain flows have been deprecated, and the three most common in use are as follows:

- **Authorization Code**: This is the recommended flow for clients with secure storage for the received token. Firstly, the client authenticates with the authorization server (on a web page such as the Google screen in *Figure 2.9*) and then receives an authorization code, which is exchanged for an access token (allowing API access), and, optionally, a refresh token (allowing the creation of a new access token). The access token can be time-limited to minimize leakage vulnerabilities, by requiring clients to re-authenticate or use the refresh token to get another access token. Both the access and refresh tokens are attractive targets for an attacker and should be stored securely on the client.

- **Authorization Code with Proof Key for Code Exchange (PKCE)**: This flow seeks to eliminate a weakness in the *authorization code* flow, namely that an attacker can intercept the code and exchange this for a token. To achieve this, a code verifier challenge is added to ensure that attackers intercepting the code cannot exchange this for a valid token. This flow is recommended for native applications (mobile applications typically) and single-page applications that are unable to store client secrets.

- **Device Authorization Flow**: This grant type is used by devices that are input-constrained (lacking a keyboard), such as smart TVs or other appliances.

- **Client Credentials Grant Type**: This flow is typically used for non-human scenarios where an approval step is impossible, such as automated processes or microservices. In this scenario, the application is authenticated using its client ID and secret.

The Gmail/Dropbox example demonstrates the *authorization code* flow, shown diagrammatically here:

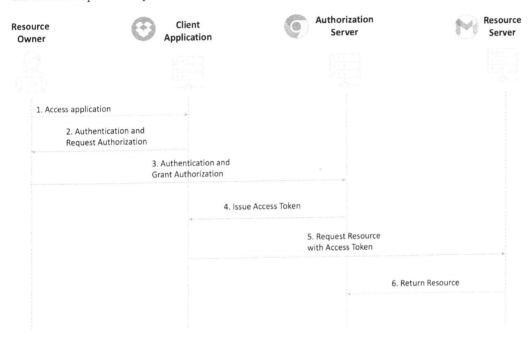

Figure 2.11 – Authorization code flow sequence

There is much more to OAuth 2 than I could cover in this section, but at this stage, ensure that you understand the basics of how the authorization code flow works and what problems OAuth2 solves for developers. In *Part 3, Defending APIs*, we will dive deeper into OAuth2 implementation and best practices.

> **Mastering OAuth2**
>
> From my own personal experience, OAuth2 can be a daunting topic to learn about, due largely to the terminology and flows. My initial perception was that it was unnecessarily complex and that I would be better off using one of the simpler methods.
>
> My recommendation is to use one of the many excellent online resources (Google or OAuth. com) and to experiment with the various flow types using an API tool such as Postman. Pause between transactions and examine the payloads (inspect the **JSON Web Tokens** (**JWTs**), for example) until you are comfortable with the protocol.

Access control best practices and methods

The good news is that the first piece of the access control puzzle, namely authentication, has been addressed by using keys and tokens (as opposed to usernames and passwords protected only by encoding in transit). The biggest challenge that remains is the distribution and secure storage of keys and tokens. To a large extent, the distribution challenge can be addressed by adopting OAuth2 and choosing the grant type most suitable for your client devices. The storage and, more broadly speaking, the lifecycle management of keys and tokens should be addressed by having a robust credential revocation and renewal process in place. Assume that credentials will be leaked, and proactively monitor for leaked credentials or indications of compromise.

The harder challenge relating to access control is that of authorization—there are far too many examples of API breaches where the root cause is poorly implemented authorization, either at the object level or function level. The root cause of many of these issues is a failure to provide granular access checks and controls—all too often, if a user is authenticated, they are allowed access to perform any action they wish. The annals of API security history are full of examples of breaches where an authenticated user was allowed to perform actions (such as deleting users or records) beyond their level of privilege.

Prevention is simultaneously simple and complicated—*always explicitly verify the client's rights to access the target object or function*. Certainly, never trust any client-supplied input values specifying the target; that is, if an authenticated client supplies an object identifier, then explicitly verify that the user belonging to that client session does have access to the object. Do not trust the session parameters or state in a cookie, as these may have been tampered with by an attacker. As a reader new to API security, you might take it that the preceding sentences state the obvious—of course, access rights should be verified. In the next chapter, *Understanding the Most Common Vulnerabilities*, you will learn that this vulnerability (called **broken object-level authorization** or simply **BOLA**) is the number one API vulnerability type.

Similarly, API developers must ensure that a client can use a function endpoint. A typical example is an administrative endpoint (say, for example, `/reset`), which should only be accessible to privileged users. Due to the nature of REST APIs, anyone who can guess this URL can call the endpoint—it is up to the backend to verify that the client has the appropriate permissions. Failure to do this is another common API vulnerability, namely **broken function-level authorization (BFLA)**.

There are two top recommendations for ensuring consistent authorization across your APIs. Firstly ensure that every API endpoint has an explicit authorization decorator or specification. Beware of implicit defaults, which may not be obvious and may change in later versions of your framework. Also, using explicit and uniform specifications will improve code reviews and allow grepping tools to verify the access controls. The following example shows a Node.js Express API endpoint with explicit anonymous access—the developer's intent is perfectly clear in this example:

```
app.get('/tasks/:taskId', auth.allowPublicAccess(), function(req,
res) {
    Task.findById(req.params.taskId, function(err, task) {
    if (err)
```

```
        res.send(err);

    res.json(task);
    });
};
```

The second recommendation is to use a standard library or framework to centrally handle your API authorization. Avoid the temptation to write your own authorization functionality since this is inefficient, lacks extensibility, is hard to maintain, and is prone to error. Much like cryptography libraries, leverage the wisdom and experience of domain experts and pick a popular implementation instead. We will cover this topic in greater detail in *Part 3*, *Defending APIs*; however, the following projects are worth investigation: Cerbos (`https://cerbos.dev/`), Casbin (`https://casbin.org/`), Keycloak (`https://www.keycloak.org/`), and Open Policy Agent (`https://www.openpolicyagent.org/`).

Using JWTs for claims and identity

In API access control, **JWTs** are used to transfer information between the client and server in a portable and robust manner. A JWT is cryptographically secure, allowing a client to verify the integrity of the message using public-key cryptography. The JSON format allows for easy transmission as part of the request header or body.

A JWT comprises three parts: the header, the claim, and the signature (**hash-based message authentication code** or simply **HMAC**). Each part is separated by a `.` character and encoded with `Base64Url` as shown:

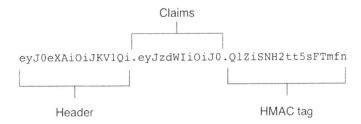

Figure 2.12 – JWT example

Let's look at these three parts in some more detail:

- The **header** provides basic metadata regarding the JWT, typically as shown here:

```
{
  "alg": "HS256",
  "typ": "JWT"
}
```

- The **claims** section contains server-specific data (the claims) in a key-value pair notation. **Registered claims** include the following:

  - `iss`: Issuer

  - `exp`: Expiration time

  - `sub`: Subject

  - `aud`: Audience

  - `nbf`: Not before

  - `iat`: Issued at

  These are not mandatory but are highly recommended.

  The **public claims** section contains specific records (defined in the IANA JSON Web Token Registry), such as client roles, groups, permissions, identity, and so on. A typical claims section is shown here:

  ```
  {
    "sid": "2df378d8-99b5-1324-c4ce-eb7574b2985a",
    "email": "colin.domoney@iloveapis.com",
    "name": "Colin Domoney",
    "role": "root"
  }
  ```

  In this example, the claim is asserting that the user identified by the email address `colin.domoney@iloveapis.com` has the role of `root` on the target system.

- The **signature** is a hash of the `Base64Url` versions of the header and the claims and a server-side secret. By verifying the signature, a client can verify that the message body was not altered or tampered with, either in transit or storage. Additionally, it can verify the identity of the signatory if used with a private/public key combination. Once the signature has been validated, the JWT claims should be fully validated to ensure the validity of registered claims (in particular, the issuer and expiration dates).

In *Part 3*, *Defending APIs*, we cover the finer details of validating JWT tokens in API backends in various popular programming languages. In conclusion, there are some key takeaways regarding JWTs:

- **Token expiration and revocation**: Since an attacker can steal a token, system designers should use short-lived tokens to minimize the impact of token leakage. Always ensure that tokens can be revoked in the backend system.

- **JWTs are used as users' digital passports**: A common use of JWTs is to store the identity and metadata of a user (as per the preceding example), much the same way as a physical password does. This means that if a token is stolen, there is no way to distinguish a rogue user from the legitimate owner. Ensure that tokens are stored securely and only transmitted via secure transport.

- **Beware of client-side storage anti-patterns**: A common anti-pattern is to store tokens in the browser's **Local Storage** because it is convenient. Unfortunately, this is vulnerable to **Cross-Site Scripting (XSS)** attacks. Using `HttpOnly` cookies minimizes the XSS attack vector but is still vulnerable to **Cross-Site Request Forgery (CSRF)** attacks.

- **Attackers can read my claims**: This is a common misconception—JWTs do not provide confidentiality of the payload (it is only encoded with `Base64Url`); however, they do guarantee integrity. A receiver can be assured a JWT has not been altered if the signature verification succeeds.

JWTs are the workhorse of modern API design, and it is essential that API defenders have a solid understanding of how they are generated and, in particular, how to handle them correctly. Before moving on to the next chapter, I would suggest using the popular `https://jwt.io/` website to experiment with some JWTs and understand how they work.

Summary

At this point, you understand all the elements of how an API is constructed. We looked at the different types of APIs in common use today and then focused on the specifics of the REST API, namely the HTTP protocol. API security is built on two foundational elements of access control, namely *authentication* (who is using the API) and *authorization* (what are they allowed to do with the API). From the origins of simple username/password authentication, we arrived at the OAuth2 framework, which is the workhorse of API access control.

Most API security vulnerabilities have their origins in poorly implemented authentication and authorization controls. We covered a number of high-level recommendations for avoiding these common pitfalls. Finally, we looked at the humble JWT, which forms the basis of API authorization, acting as your digital passport.

In the next chapter, we are going to look at how things can go wrong in practice by taking a deep dive into the most common API security vulnerabilities by looking at the OWASP API Security Top 10.

Further reading

Further reading on OAuth2 is available here:

- `https://developers.google.com/oauthplayground/`

- `https://www.oauth.com/playground/`

- `https://developer.okta.com/blog/2019/10/21/illustrated-guide-to-oauth-and-oidc`

Further reading on JWTs is available here:

- `https://arielweinberger.medium.com/json-web-token-jwt-the-only-explanation-youll-ever-need-cf53f0822f50`

- `https://jwt.io/`

Finally, further reading on authorization is available here:

- `https://learning.postman.com/docs/sending-requests/authorization/`

- `https://thenewstack.io/oso-tackles-unbundling-security-authorization/`

- `https://www.openpolicyagent.org/docs/v0.11.0/http-api-authorization/`

3

Understanding Common API Vulnerabilities

Now that we understand how APIs are constructed, we will turn our attention to the core topic of this book—*API security*. In this chapter, we will focus on the different types of vulnerabilities that can adversely impact API security, gaining an understanding of the underlying cause, the impact, and the recommended prevention or mitigation for each.

In this chapter, we're going to cover the following main topics:

- The importance of vulnerability classification
- The Open Worldwide Application Security Project API Security Top 10 vulnerabilities
- Vulnerabilities versus abuse cases
- Business logic vulnerabilities

The importance of vulnerability classification

Security researchers have long understood the importance of classifying vulnerabilities within software and hardware systems. Classification allows researchers to group similar vulnerabilities together based on their characteristics and then apply standard patterns for mitigation and protection.

Flaws versus vulnerabilities versus exploits versus threats versus risks

The preceding terms cause confusion in the security industry, so it is worth disambiguating them as they are subtly different.

A *flaw* is an implementation defect or weakness in code that may be latent or exploitable. A *vulnerability* is a flaw that can be exploited by an attacker. An *exploit* is a procedure or method used by an attacker to take advantage of a flaw, that is, it is the "how" of a vulnerability. A *threat* is anything that has the potential to do harm to a system, and it can be intentional (a hacker) or unintentional (forgetting to patch a system) in nature. Finally, the *risk* is the product of the vulnerability rating, the level of threat, and the impact: *Risk = Vulnerability x Threat x Impact.* The following diagram illustrates the core concept well: in quadrant I, there is both a high impact and probability, and hence a high risk. In quadrant III, there is a low risk of both, and hence the risk is low.

The most widely used classification system is the **Common Weakness Enumeration** (CWE) system operation by the Mitre Corporation (`https://cwe.mitre.org/index.html`). The CWE defines 927 different classes of vulnerability and ranks them by the severity of impact; the top 25 software vulnerabilities are shown in the following table:

Rank	ID	Name	Score	KEV Count (CVEs)	Rank Change vs. 2021
1	CWE-787	Out-of-bounds Write	64.20	62	0
2	CWE-79	Improper Neutralization of Input During Web Page Generation ('Cross-site Scripting')	45.97	2	0
3	CWE-89	Improper Neutralization of Special Elements used in an SQL Command ('SQL Injection')	22.11	7	+3 ▲
4	CWE-20	Improper Input Validation	20.63	20	0
5	CWE-125	Out-of-bounds Read	17.67	1	-2 ▼
6	CWE-78	Improper Neutralization of Special Elements used in an OS Command ('OS Command Injection')	17.53	32	-1 ▼
7	CWE-416	Use After Free	15.50	28	0
8	CWE-22	Improper Limitation of a Pathname to a Restricted Directory ('Path Traversal')	14.08	19	0
9	CWE-352	Cross-Site Request Forgery (CSRF)	11.53	1	0
10	CWE-434	Unrestricted Upload of File with Dangerous Type	9.56	6	0
11	CWE-476	NULL Pointer Dereference	7.15	0	+4 ▲
12	CWE-502	Deserialization of Untrusted Data	6.68	7	+1 ▲
13	CWE-190	Integer Overflow or Wraparound	6.53	2	-1 ▼
14	CWE-287	Improper Authentication	6.35	4	0
15	CWE-798	Use of Hard-coded Credentials	5.66	0	+1 ▲
16	CWE-862	Missing Authorization	5.53	1	+2 ▲
17	CWE-77	Improper Neutralization of Special Elements used in a Command ('Command Injection')	5.42	5	+8 ▲
18	CWE-306	Missing Authentication for Critical Function	5.15	6	-7 ▼
19	CWE-119	Improper Restriction of Operations within the Bounds of a Memory Buffer	4.85	6	-2 ▼
20	CWE-276	Incorrect Default Permissions	4.84	0	-1 ▼
21	CWE-918	Server-Side Request Forgery (SSRF)	4.27	8	+3 ▲
22	CWE-362	Concurrent Execution using Shared Resource with Improper Synchronization ('Race Condition')	3.57	6	+11 ▲
23	CWE-400	Uncontrolled Resource Consumption	3.56	2	+4 ▲
24	CWE-611	Improper Restriction of XML External Entity Reference	3.38	0	-1 ▼
25	CWE-94	Improper Control of Generation of Code ('Code Injection')	3.32	4	+3 ▲

Figure 3.1 – Top 25 software vulnerabilities

Taking the example of SQL Injection (CWE-89), a vulnerability occurring in APIs, the CWE entry shows the impact of the vulnerability according to the **Confidentiality**, **Integrity**, and **Availability** (**CIA**) triad:

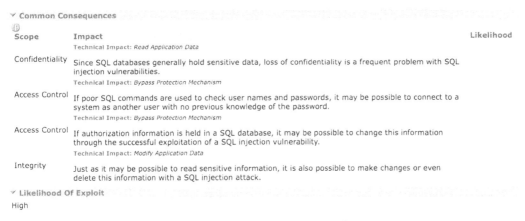

Figure 3.2 – SQL injection CIA triad

The CWE entry also provides references to real-world examples of this vulnerability by referencing their **Common Vulnerabilities and Exposures (CVE)** identifier:

Figure 3.3 – Example vulnerabilities for CWE-89

The CWE entry also contains useful information about how the vulnerability can be detected using techniques such as static and dynamic analysis:

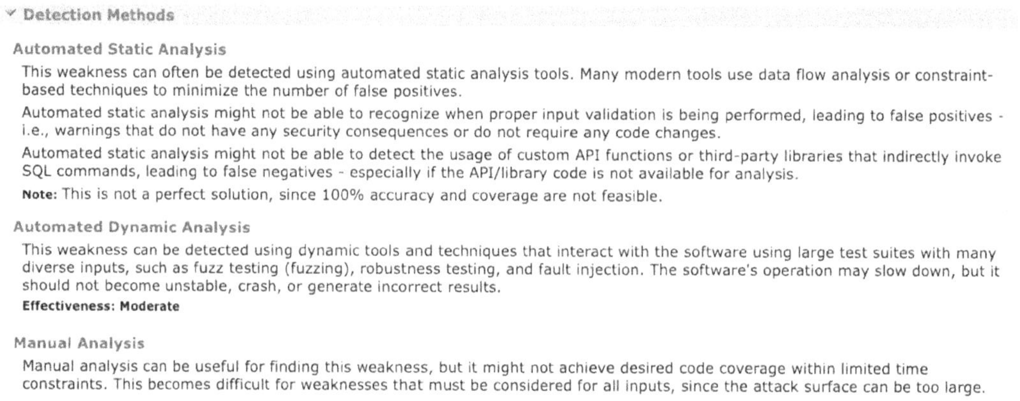

Detection Methods

Automated Static Analysis

This weakness can often be detected using automated static analysis tools. Many modern tools use data flow analysis or constraint-based techniques to minimize the number of false positives.

Automated static analysis might not be able to recognize when proper input validation is being performed, leading to false positives - i.e., warnings that do not have any security consequences or do not require any code changes.

Automated static analysis might not be able to detect the usage of custom API functions or third-party libraries that indirectly invoke SQL commands, leading to false negatives - especially if the API/library code is not available for analysis.

Note: This is not a perfect solution, since 100% accuracy and coverage are not feasible.

Automated Dynamic Analysis

This weakness can be detected using dynamic tools and techniques that interact with the software using large test suites with many diverse inputs, such as fuzz testing (fuzzing), robustness testing, and fault injection. The software's operation may slow down, but it should not become unstable, crash, or generate incorrect results.

Effectiveness: Moderate

Manual Analysis

Manual analysis can be useful for finding this weakness, but it might not achieve desired code coverage within limited time constraints. This becomes difficult for weaknesses that must be considered for all inputs, since the attack surface can be too large.

Figure 3.4 – Detection methods for CWE-89

The advantages of using a classification scheme are as follows:

- Allows risks to be grouped together by their fundamental characteristics

- Allows harmonization among various security tools, that is, everyone is using the same language

- Allows common defense and protection techniques to be employed, for example, for SQL injection, there is a standard list of recommendations for remediation in the form of a *cheat sheet* produced by the Open Worldwide Application Security Project.

The primary advantage is that CWEs allow *software vulnerabilities* to be translated to the *security domain*—a software flaw can be assigned a risk rating based on its impact on the CIA triad.

CVEs and CWEs—two very different things

Beware of confusing these two similar acronyms. CVEs refer to an instance of a vulnerability that was exploited and/or disclosed (that is, the Equifax Apache Struts vulnerability, referred to as CVE-2017-5638). The CWE is a classification scheme for describing a particular vulnerability type (in the Equifax example, the vulnerability type was CWE-94: Improper Control of Generation of Code ('Code Injection')).

CVEs refer to a specific occurrence of a vulnerability and may include one or more CWE types.

Now that we understand how vulnerabilities are classified, let's look at the all-important Open Worldwide Application Security Project API Security Top 10, which forms the foundation of the rest of this chapter. The **Open Worldwide Application Security Project** is a global community of application security experts who advocate the need for improved software security via projects, meetups, guides, and standards.

Its highest profile publication is the Open Worldwide Application Security Project Top 10, which ranks the highest impact vulnerability classes based on their impact on an application. Its scoring method considers several factors to produce an average risk score, as shown:

Threat Agents	Exploitability	Weakness Prevalence	Weakness detectability	Technical impacts	Business Impacts
Application-specific	Easy: 3	Widespread: 3	Easy: 3	Severe: 3	Business-specific
	Average: 2	Common: 2	Average: 2	Moderate: 2	
	Difficult: 1	Uncommon: 1	Difficult: 1	Minor: 1	

Figure 3.5 – Open Worldwide Application Security Project Top 10 scoring methodology

The Open Worldwide Application Security Project Top 10 is a tremendously useful indicator for security and development teams as it provides an instant view of the most impactful security vulnerabilities. This allows teams to focus on the issues that matter most, namely the ones that will cause the most damage to their organizations. Having said that, you should avoid the fallacy that lower-ranking issues can be ignored. This is not the case; they can be de-prioritized in favor of more critical issues, but they do still need to be addressed.

The Open Worldwide Application Security Project Top 10 focuses primarily on web applications (given the origins of the Open Worldwide Application Security Project); however, the Open Worldwide Application Security Project decided that APIs present their own unique vulnerabilities. This led to the introduction of the Open Worldwide Application Security Project API Security Top 10, which will be our focus for this chapter. The top 10 is updated periodically based on emerging incident data; the next update to the Open Worldwide Application Security Project API Security Top 10 is due in 2023. In this book, we will refer to the 2019 version, although we will take a brief look at the breaking changes in the 2023 version at the end of this chapter.

> **Don't be overly reliant on the Open Worldwide Application Security Project API Security Top 10**
>
> While the top 10 serves an incredibly useful purpose in focusing attention and effort, you should avoid being too narrowly focused on the top 10 and consider a broader spectrum of API risk.
>
> For example, APIs with business logic vulnerabilities may have a massive impact, as can APIs using vulnerable libraries or components. Start with the top 10 but keep an eye out for other risk types.

This section has provided a formal understanding of much of the nomenclature associated with vulnerabilities and how they relate to risk. The key consideration is that security work should begin with the highest risk items and proceed, as time and budget permit, to lower risk items.

Exploring the Open Worldwide Application Security Project API Security Top 10

Let us now start our exploration of the Open Worldwide Application Security Project API Security Top 10 vulnerabilities. Although the standard Open Worldwide Application Security Project listing provides the vulnerabilities in decreasing order of severity, I have chosen to group them by vulnerability type and root cause to aid understanding. Shall we begin?

Object-level vulnerabilities

There is only one object-level vulnerability, which is the now infamous broken object-level authorization, which is number one in the Open Worldwide Application Security Project API Security Top 10.

API1:2019—Broken object-level authorization

The easiest real-world analogy to understand **broken object-level authorization** (**BOLA**) is that of a coat check-in at an entertainment venue. Upon arrival, you drop your coat off with the clerk and are given a ticket with a number, let's say #10, for example. Now when you go to check out your coat, you change the number to #70, and the clerk goes and retrieves the coat with #70 instead of your original coat. The vulnerability is that the processor (the clerk) failed to check whether the client (you) had access to the object (the new coat). That's BOLA in a nutshell.

Let's get into the technical details by using the following diagram:

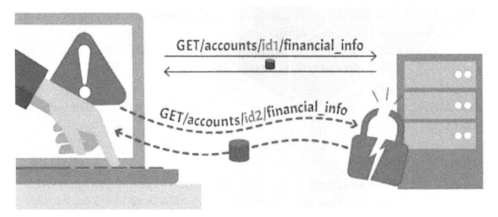

Figure 3.6 – BOLA

With BOLA, an endpoint allows a given user access to an object (namely data) that does not belong to that user. The root cause of this vulnerability is a failure in the API backend to validate that the requesting client has access to the specified object. Typically, attackers will gain authentication to an API and then attempt to manipulate object identifiers to probe for poor implementations.

In this example, the user authenticated themselves and was given access to their financial information using identifier id1. They then made a subsequent call to the API using their existing session but switched the requested identifier to id2. The backend should have rejected this request, but, as in many cases, it simply processed the request given that the request belonged to a valid session.

The prevention of BOLA is conceptually simple but practically very difficult—always fully validate the permissions of a client to access a given object.

The recommended defense against BOLA is to use an **authorization decision engine** to authorize *each and every* access to an object. This is most easily understood by a code example showing an incorrect and then a correct implementation.

In this code snippet, the controller fails to check whether the caller has access to the object specified by the id parameter and returns the object via a lookup:

```
Class UserController < ApplicationController
  def show
    @this_user = User.find(params[:id])
    render json: @this_user
  end
end
```

Now, by adding an authorization decision engine to the code, the controller validates that the session user (current_user) has permission to access the specified receipt ID:

```
Class UserController < ApplicationController
  def show
    if Autorization.user_has_access(current_user,
      params[:id])
        @this_user = User.find(params[:id])
        render json: @this_user
  end
end
```

There are various partial mitigations that can be applied to BOLA, including using random UIDs to prevent easy guessing; using opaque, temporary IDs; and using IDs in the JWT token. None of these are sufficient to prevent BOLA and will merely serve to delay an attacker.

That is a whistle-stop tour of BOLA; we're going to be seeing a lot more of this pernicious vulnerability throughout the rest of this book.

Further information is available at https://apisecurity.io/encyclopedia/content/ owasp/api1-broken-object-level-authorization and https://inonst. medium.com/a-deep-dive-on-the-most-critical-api-vulnerability-bola- 1342224ec3f2.

Authentication vulnerabilities

Broken or missing authentication is a leading cause of API vulnerability. As described in *Chapter 2, Understanding APIs*, a wide range of authentication mechanisms exist, and often, developers are overwhelmed when choosing and implementing a method, leading to exploits in production. The Open Worldwide Application Security Project has rated `API2:2019—Broken authentication` as the second most critical API vulnerability.

API2:2019—Broken authentication

From my personal experience, broken authentication is the most frequently occurring API vulnerability and certainly can have catastrophic consequences. There are two variants:

- **Missing authentication**: As surprising as this may seem, many real-world breaches show a total lack of authentication—the developers made no attempt to implement authentication at all.

- **Broken authentication**: This is a harder problem to solve—often, developers will implement an authorization mechanism but it turns out to be insecure.

The reasons for broken authentication are numerous and include the following:

- Weak API keys that are not rotated

- Weak (or blank) passwords that can easily be guessed using brute-force attacks

- Poor access token validation (JWTs are frequently validated incorrectly)

- Credentials and keys are included in URLs

- Failure to use industry best practice in authentication backends

- Abuse of password reset mechanisms

- Poor cryptographic practices, including the use of weak ciphers, insufficient entropy, and short keys

For every API endpoint, be sure to enforce authentication, ideally via built-in middleware or approved and validated mechanisms. Missing authentication can be detected easily by scanning both API specifications and source code; however, broken authentication is harder to detect and relies on skilled security analysts.

Further information is available here: `https://apisecurity.io/encyclopedia/content/owasp/api2-broken-authentication`.

Function-level vulnerabilities

Although only ranking fifth in the Open Worldwide Application Security Project's top 10, function-level vulnerabilities are potentially very serious, especially in relation to administrative privilege-level operations. Let's look at broken function-level authorization in detail.

API5:2019—Broken function-level authorization

While BOLA, as covered earlier, concerns unauthorized access to an object (typically data), broken function-level authorization is concerned with unauthorized access to an API function or endpoint. Typically, this occurs in one of two ways: the API provides an undocumented endpoint intended for debugging or management that lacks authentication, or an endpoint that is authenticated but is intended to be restricted to specific users, not everyone.

A typical example is an API that has an /admin endpoint intended for use by a higher-privilege user. In a vulnerable implementation, the API backend code fails to validate that the session user is authorized to access the endpoint—it is not sufficient to only check that the user is authenticated.

This is best illustrated visually:

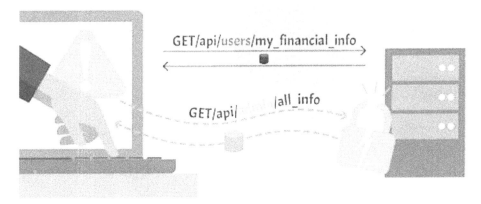

Figure 3.7 – Broken function-level authorization

In this example, a regular user has access to an /api/users/my_financial_info endpoint but discovers by trial and error that an /api/admin/all_info endpoint exists and that they also have access to this endpoint. The endpoint has failed to validate that a regular user can access a privileged endpoint.

In the same way that BOLA should fully validate access to the requested object, it is essential that an endpoint fully validates access to the requested function. Usually, this is achieved by means of claims within JWTs or some implementation of RBAC. The golden rule is to ensure that although a user is authenticated, this does not imply they have open access to all functions. For high-privilege endpoints, apply the principle of least privilege or deny by default.

Further information is available here: `https://apisecurity.io/encyclopedia/content/owasp/api5-broken-function-level-authorization`.

Data vulnerabilities

APIs are predominantly responsible for processing data from a client to a server or vice versa. It should come as no great surprise that data vulnerabilities are among the most severe for APIs. There are two vulnerabilities in this category: excessive data exposure (affecting response data) and mass assignment (affecting request data). Both carry a significant risk impact on APIs and underlying data.

The good news is that data vulnerabilities can be easily identified and detected using conformance scanning of APIs according to their intended design in the API contract.

API3:2019—Excessive data exposure

Rated third according to the Open Worldwide Application Security Project is one of the most prevalent and harmful API vulnerabilities—excessive data exposure. In simple terms, the API returns more data than expected or is necessary to meet its purpose. For example, if a client attempts to retrieve user details, a vulnerable API method might return **Personally Identifiable Information** (**PII**) data. The API should return the exact amount of data to meet the design requirement—no more, no less.

This vulnerability category features in nearly all high-profile API breaches—this is hardly surprising since the intent behind many API attacks is data exfiltration.

Fortunately, this is one of the easiest vulnerability categories to eliminate. The key is the adoption of the OpenAPI Specification to define the data contract of the API—in this case, the response data body. Once the contract is defined, it can be used to enforce granular control at layer 7 of the API using dedicated API firewalls. But this is reliant on a full understanding of the data objects to be transferred via the API—this is where API governance plays a vital role.

You may be wondering why this vulnerability category is so prevalent — here are a few reasons:

- API developers may try and be "*helpful*" to consumers by returning more data than necessary, just in case it is needed in future releases. They are reliant on client-side filtering, which can obviously be subverted.

- In many cases, data access controls are neither defined nor enforced. Do not rely on developers to understand data sensitivity; make sure you have a solid data governance process in place.

- In most cases, however, it comes down to a case of developer efficiency. Modern languages provide powerful **Object Relational Mappers** (**ORMs**), which abstract database access from the developer—the database becomes a collection of strongly typed objects, and the ORM handles the database interface. Fortunately for developers (and unfortunately for security teams), it is a trivial matter to convert a complex data object into a JSON API response by invoking the `to_json()` method on the object. It's all too easy to inadvertently expose your entire data object via an API.

The problem is best visualized with a real-world example. In this case, the database stored a plain-text representation of the user password (a serious issue in its own right), and this is returned in a listing of all users, as shown:

GET   ⌄   {{42c url}}/api/admin/all_users

Params   Authorization   Headers (7)   Body   Pre-request Script   Tests   Settings

Body   Cookies   Headers (7)   Test Results

Pretty   Raw   Preview   Visualize   JSON ⌄   ⇆

```
1    [
2        {
3            "_id": 2,
4            "email": "misty94@hotmail.com",
5            "password": "ball",
6            "name": "Brenna Lehner",
7            "pic": "https://s3.amazonaws.com/uifaces/faces/twitter/herrhaase/128.jpg",
8            "is_admin": false,
9            "account_balance": 48.350000000000094
10       },
11       {
12           "_id": 1,
13           "email": "beth.white@gmail.com",
14           "password": "mouse",
15           "name": "Savion Emmerich II",
16           "pic": "https://s3.amazonaws.com/uifaces/faces/twitter/markolschesky/128.jpg",
17           "is_admin": false,
18           "account_balance": 48.350000000000094
19       },
20       {
21           "_id": 3,
22           "email": "garrison.buckridge@yahoo.com",
23           "password": "keyboard",
24           "name": "Eric Veum",
25           "pic": "https://s3.amazonaws.com/uifaces/faces/twitter/lepinski/128.jpg",
26           "is_admin": false,
27           "account_balance": 48.350000000000094
28       },
```

Figure 3.8 – Example of excessive data exposure

Correct implementation of this API endpoint would have removed the password field rather than returning it or perhaps obfuscating the password by masking it in the API backend.

The most important consideration for protecting against excessive information exposure is never to rely on client-side protections, such as masking sensitive data in the browser or mobile application or avoiding displaying the information at all. Remember, once the API has responded with the data, it can be intercepted either in transit (via a machine-in-the-middle attack) or within the client (by using browser tools or debuggers). As we will see in the next chapter, an over-reliance on client-side protections is one of the leading causes of information leakage in API breaches.

As a developer, the key to preventing excessive data exposure is to have a clear understanding of the following:

- The *minimum set of data* required by the consumer of the API in question. Do not return more data on the off chance the consumer may need it in the future—in this event, update the API to a new revision on demand.

- The *data privacy and protection requirements* of the data your API processes—if the data is PII, financial or medical records, or other sensitive data, strict access controls must be used to avoid unwanted exposure. Check with your **Data Protection Officer** (**DPO**) or governance and risk teams if unsure—do not assume.

By using accurately specified OpenAPI definitions, it is possible to protect APIs against excessive data exposure at runtime.

Further information is available here: `https://apisecurity.io/encyclopedia/content/owasp/api3-excessive-data-exposure`.

API6:2019—Mass assignment

The second type of data vulnerability is mass assignment and is almost the inverse of excessive data exposure—rather than returning too much data, a mass assignment vulnerability occurs when an API accepts too much data in the request.

Mass assignment occurs when the API backend interprets additional data supplied by the requestor. Commonly, API backends will use features in their API framework that allow them to assign input data to a backend object in a single operation as an efficiency aid—hence the term mass assignment. While in some cases this may be helpful, it can cause unexpected results if an attacker is able to guess one of the fields and inject a matching value. A typical example is when an attacker guesses a field name (such as `is_admin`) and assigns this to `true`. A properly implemented backend would ignore such extraneous fields. If, however, a mass assignment function was used in the backend, this injected value would be interpreted and persisted to the API's underlying data store.

This vulnerability is shown here:

Figure 3.9 – Example of mass assignment

In this example, the API backend is expecting a JSON object with email and name fields, but an is_admin field as provided. Unfortunately, in this case, the backend does not check for this unexpected field and commits the field to the database (normally using an ORM), effectively elevating the user's privilege in this example.

An attacker can either guess the names of fields, use a **fuzzer** (a software tool capable of generating synthetic data to be used on a target for testing or attack purposes) for this purpose, or try and reverse-engineer field names from the documentation or via some of the GET requests.

Mass assignment is relatively easy to defend against—avoid automatically binding input fields (which are untrusted) to internal data objects. Often, sensitive fields can be marked as read-only, which allows them to be read via an API but prevents updating them via the API.

By using accurately specified OpenAPI definitions, it is possible to protect APIs against mass assignment at runtime.

Further information is available here: https://apisecurity.io/encyclopedia/content/owasp/api6-mass-assignment.

Configuration vulnerabilities

This rather broad category covers three Open Worldwide Application Security Project Top 10 vulnerabilities affecting the configuration of APIs:

- Lack of resources and rate limiting

- Security misconfiguration

- Improper asset management

Although all vulnerabilities warrant due care and attention, the good news is that this class of vulnerability can usually be prevented using API gateways and firewalls, with scanners and fuzzers, or through inspection.

API4:2019—Lack of resources and rate limiting

One of the easiest and least subtle attacks against an API is a **Denial of Service (DoS)** attack, whereby an attacker will submit high volumes of requests to an API endpoint until the API fails due to resource exhaustion. Even if the API does not fail totally, such an attack may lead to an increase in resource usage, leading to increased costs.

Brute-force attacks can also be used to defeat password reset mechanisms or login endpoints using credential stuffing attacks. Another scenario where brute-forcing attacks work well is against APIs taking parameters in a standard format, for example, courier tracking numbers.

For an API business, consideration must be given to their fair use policy and/or their pricing tiers for API access. While quota enforcement (a business concern) is subtly different from rate-limiting (a security concern), they are related if a DoS attack is able to limit access to a premium-tier customer, causing customer dissatisfaction. Robust rate-limiting is important both for security and for usability and customer experience.

Rate-limiting strategies can be either a *hard stop* where the API will respond with **429 too many requests** until a timeout expires or a *throttled stop* where the API implements a delay in responding to requests, causing the effective throughput to be limited.

Rate-limiting can be implemented in various places:

- **In the API backend code itself**: This moves the onus to developers, and this can be difficult to implement reliably.

- **In an API gateway**: This is the standard implementation point as external APIs will be routed via the gateway, which understands API traffic and is well suited to performing hard-stop limiting before the API backend is even called.

- **In a dedicated API firewall, which enforces the OAS contract on the API backend**: The contract can be decorated with vendor definitions specifying the rate-limiting parameters. Again, the limiting is enforced before the API backend is called.

While rate-limiting is essential, be aware that skilled attackers may be able to defeat it by spoofing IP addresses, using a distributed attack, changing clients or user agents, or switching parameters.

Further information is available at `https://apisecurity.io/encyclopedia/content/owasp/api4-lack-of-resources-and-rate-limiting` and `https://theauthapi.com/articles/right-ways-of-rate-limiting/`.

API7:2019—Security misconfiguration

Security misconfiguration covers a wide range of operational or runtime misconfigurations, which result in vulnerable infrastructure for an API. These can include issues relating to misconfigured headers, missing transit encryption, accepting unused HTTP methods, overly verbose error messages, and lack of sanitization.

The first important topic to cover is that of security headers, which are used by HTTP servers to instruct the client on how to act on the response returned. Headers can be added by API servers to provide metadata useful for metrics or tracing. In the hands of an attacker, these additional headers can be used against the API. For example, the `X-Powered-By` header indicates the server backend, which is useful information for an attacker, who can search for published vulnerabilities. The `X-Content-Type-Options: nosniff` header prevents the client from making assumptions about the response, although it is a best practice to explicitly specify the content type with `Content-Type: application/json`. The `X-XSS-Protection header` indicates the type of **Cross-Site Scripting** (**XSS**) used; a value of 0 indicates none, which is useful to an attacker. Another header of interest is `X-Response-Time`, which is the server-side processing time. In the hands of an attacker, such information allows them to launch sophisticated **timing attacks** to perform discovery on the API implementation.

The good news is that excellent guidance on security headers is offered by the Open Worldwide Application Security Project and others, and the correct header implementation can be tested using online scanners.

Another very simple category to eliminate is the exposure of unnecessary services, features, and interfaces. Audit your server implementation and harden it to disable anything not necessary for the API consumers. Typically, this includes things such as hidden debug and administration interfaces, which can be abused.

Many API backend servers or frameworks include default accounts with well-known passwords for debugging or initialization use—ensure that these are disabled or that the passwords are at least changed.

API frameworks (such as .NET Core or Django) will implement default API endpoints and behaviors to increase developer efficiency. Unless the developer is aware of these built-in defaults, it is possible they will inadvertently deploy APIs with HTTP methods with default implementations that an attacker can abuse. For example, a GET method may be defined; however, an attacker may attempt to POST data to this endpoint. Remove unused methods and scan the API against the OpenAPI contract.

Finally, a lack of sanitization can allow attackers to upload malicious content (such as corrupted ZIP or PDF files, or even Trojans or malware). Never trust user-supplied input.

Further information is available here: `https://apisecurity.io/encyclopedia/content/owasp/api7-security-misconfiguration`.

API9:2019—Improper asset management

Improper asset management occurs when an organization exercises inadequate governance over its API development process. Modern development practices have made it increasingly easy for development teams to develop and deploy APIs, often hidden from the sight of their security teams. This inevitably leads to *API sprawl* with multiple versions of APIs deployed in multiple regions or environments. Older versions are infrequently maintained but may still provide access to production data.

This results in the dual problems of *shadow APIs* (APIs invisible to the security or governance teams) and *zombie APIs* (deprecated, obsolete, or unmaintained APIs still in operation).

Unfortunately, this is one of the hardest categories of vulnerability to address, as its root cause originates in the organization's culture and processes rather than due to technology failings. If a development team or a business unit is under pressure to deliver on business outcomes and is hamstrung by an overly rigorous process to deploy new applications, they may resort to using shadow IT (IT deployed without the knowledge of internal IT, security, and governance teams directly by a product team, typically by directly purchasing cloud resources) to deploy APIs. This gives rise to *shadow APIs* that are outside of normal security and governance processes. Similar *zombie APIs* are created when a development team or business unit lacks the sufficient budget to keep an API update in the light of required changes (for security updates or functionality updates) and opts to keep an outdated API in service. These pose a security risk to an organization since they may utilize vulnerable libraries and components or contain implementation weaknesses.

A robust API governance process is key to addressing such issues—senior leadership should work with stakeholders across the organization to define an API governance process, and then this should be adhered to across all teams. Legacy or technical debt should be addressed as and when time permits but should not be ignored.

Further information is available here: `https://apisecurity.io/encyclopedia/content/owasp/api9-improper-assets-management`.

Implementation vulnerabilities

Finally, this catch-all category of vulnerabilities covers various implementation and coding vulnerabilities, including issues such as input injection (SQL, command, LDAP, etc.), and a lack of logging and monitoring to detect API attacks or breaches.

API8:2019—Injection

SQL injection attacks have existed for over two decades now and are still one of the most prevalent attacks on web applications. Unfortunately, attackers have rapidly turned their focus to APIs using the same techniques to exploit API vulnerabilities. APIs provide several opportunities for injection, either in API parameters or in the payload of the request body.

Other types of injection attacks include NoSQL, LDAP, OS command injection, and XML parser injection.

Injection attacks are still notoriously prevalent despite their infamy in the security industry. This is in large part due to the relative complexity of mitigating this type of attack using a combination of filters, sanitizers, parsers, and validators. While these methods may prove effective, they are often poorly implemented or can be subverted by skilled attackers.

For the particular case of SQL injection, the foolproof prevention is to use parameterized database queries (as opposed to using string interpolated queries that are vulnerable to injection). However, in many cases, this requires large-scale code refactoring.

Further information is available here: `https://apisecurity.io/encyclopedia/content/owasp/api8-injection`.

API10:2019—Insufficient logging and monitoring

Finally, the Open Worldwide Application Security Project identifies insufficient logging and monitoring as an API vulnerability. In many API breaches (discussed in the next chapter), the organization was not aware that the breach was in progress or had occurred in the first instance. For this reason, it is essential to log and monitor all API activity, either for prevention or post-mortems.

What should (and shouldn't) be logged?

- Log failed attempts, denied access, input validation failures, violations in security policy, or rate-limiting

- Include enough detail to identify attackers, such as IP addresses, usernames, and so on

- DO NOT include sensitive information (passwords, tokens, PII, account numbers, etc.) in logs

- Use a standard logging format to facilitate interoperability with other systems such as **Common Event Format (CEF)**, **Syslog**, and so on

While important for deep-dive post-mortem logs, the key to continuous and automated API security monitoring is the integration of API events into **Security Information and Event Management (SIEM)** systems and **Security Operation Centers (SOCs)**. By surfacing API-specific security events within standard security monitoring tools, it is possible for security operations teams to continuously monitor for probes, discovery, enumeration, and active attacks against APIs. This affords the luxury of being able to detect attacks in real time rather than after the fact.

Further information is available here: `https://apisecurity.io/encyclopedia/content/owasp/api10-insufficient-logging-and-monitoring`.

That concludes our exploration of the Open Worldwide Application Security Project API Security Top 10—broken authorization and authentication are top of the list, but data vulnerabilities are persistent and impactful vulnerabilities.

Vulnerabilities versus abuse cases

While the discussion so far has focused on vulnerabilities (flaws in software that can be exploited by an attacker), we need to also consider the impact of API abuse on API security. API abuse is generally defined as the use of an API in an unexpected way, leading to negative consequences. Normally, an API is designed to support a mobile application or website; however, since the API is exposed, curious users or adversaries can reverse-engineer the API and use it for their own purposes.

An excellent example comes from the supermarket industry in the United Kingdom during the first Covid-19 lockdown. Supermarket delivery services rapidly became oversubscribed, and supermarkets quickly implemented limiting controls on the web frontends to avoid total overload and failure. Curious developers quickly examined the APIs and found endpoints that allowed access to the booking system and were able to reserve precious delivery slots by bypassing the frontend controls. There was no exploit of any vulnerability in the APIs; however, due to a change in user patterns (due to the lockdown), their APIs were used in a way the designers simply could not have expected.

Similar examples include rival airlines scraping pricing information from their websites, property aggregation sites being mined to predict house pricing trends, or scraping shopping websites to find the best prices.

Detailed examples of API abuse cases and mitigations are provided next.

Exploring abuse cases

As mentioned, API abuse is an increasing risk to APIs—while abuse cases do not generally result in large-scale breaches reaching news headlines, they can inflict significant financial or reputational damage to organizations.

API abuse is the use of an API in a way that was not intended. The most common abuse cases include the following:

- **Excessive pagination**: To avoid returning excessively large response bodies, APIs allow the client to specify a page of records to return, typically via the **Next** button in a browser. However, in an API, a script could automatically page through all records, effectively downloading the entire record set. This is certainly not how the designer intended this feature to be used.

- **Data exfiltration**: Like the excessive data exposure vulnerability, data exfiltration allows a user to parse API responses, looking for sensitive or interesting data that may be traded on the black market.

- **Scrapers**: Web applications built on top of API backends can be automatically *scraped* using a UI testing framework (such as Selenium or BrowserStack) to extract data en masse. It is also possible to use web browser developer tools to directly inspect the APIs and automate scraping directly via the API.

- **Credential abuse**: Since API requests can easily be automated with scripts, they are ideal targets for credential abuse attacks using dictionaries of leaked credentials or common username/password combinations. Rate-limiting is key to defeating this abuse vector.

- **Scripts or tools**: Probably the most common form of abuse is via a virtual army of tech-savvy users using freely available scripts or tools (covered in the section on *Part 2, Attacking APIs*). Often, the intent is benign and fueled merely by curiosity; however, so-called *script-kiddies* can cause significant frustration to security teams protecting APIs.

- **Machine-in-the-middle attacks**: While the use of **Transport Layer Security** (**TLS**) has largely eliminated the risk of eavesdropping on data in transit, mobile applications present a risk vector to actors wishing to intercept data. Unless the mobile application correctly implements certificate pinning, an attacker can use a reverse proxy to intercept data between the application and its backend API.

- **DDoS attacks**: Finally, the most harmful of the abuse cases is the use of bot armies deployed to abuse APIs en masse with the intent of degrading service or taking it offline entirely. Distributed bot armies can be difficult to defeat using rate-limiting methods due to their distributed nature (different source IP addresses).

API abuse is difficult to protect against in many cases since, after all, there is no explicit vulnerability in the API; however, it is being used against the intent of the designer. Standard monitoring and defense mechanisms are ineffective in protecting against API abuse; after all, the API is behaving as intended, but being used in an unexpected way. The key to protecting against API abuse is to model standard API usage patterns and then use continuous monitoring and modern **machine learning** (**ML**) techniques to detect deviations from standard usage patterns.

The Cloudflare security team has put together an excellent guide on protecting against API abuse: `https://blog.cloudflare.com/api-abuse-detection/`.

Business logic vulnerabilities

The final category of vulnerability is that of business logic vulnerabilities, which allow an attacker to elicit unexpected behaviors with negative consequences. They are closely aligned with abuse cases and can be notoriously difficult to eliminate. The key is to think like an adversary—threat modeling exercises can be useful in highlighting business logic vulnerabilities.

There are several good examples of business logic vulnerabilities affecting APIs:

- **Relying on client-side controls**: One of the most nefarious and persistent weaknesses (highlighted frequently in this chapter) is the reliance on client-side controls to implement security. Simply put, they do not work and can always be defeated—do not use them.

- **Trusting users**: Avoid trusting that the user will behave in the way you intended; typical examples include not providing required parameters or supplying the wrong format of data.

- **Trusting partners and third parties**: The use of partner and third-party APIs opens additional risk. As a consumer, be wary of trusting data from upstream APIs—always fully validate. As a supplier, avoid relying on your consumers to be trustworthy—they may use your API in an egregious fashion (such as exposing it to the internet).

- **Trusting user-supplied data**: This is the number one bane of security teams—user input can be malformed accidentally (submitting an invalid ZIP file) or deliberately formulated to elicit unexpected behavior in the consuming application. This is the root cause of all the injection attacks covered earlier.

- **Relying on an intended sequence**: APIs are often designed to support a mobile application that behaves in a very predictable manner (the calls are always in the same sequence). It is very easy for an attacker to observe the API requests and sequencing, and then attempt to use the API calls in a different sequence to produce unexpected behavior. Typically, attackers can use this technique to bypass validation or authentication steps.

- **Expecting users to follow instructions**: Finally, despite the clearest and most concise instructions for API usage, it is unlikely that users will follow these instructions precisely. Be sure that APIs defend against misoperation by users, and in particular, use fuzzing and random tests to identify assumptions about user behavior.

Business logic flaws are hard to test because of the wide combination of behaviors, both expected and unexpected. Learn to think like an attacker and avoid making assumptions about your users or environments.

Preview of the Open Worldwide Application Security Project API Security Top 10 2023

As mentioned earlier in this chapter, the Open Worldwide Application Security Project API Security Top 10 is undergoing changes to reflect the API threat landscape in 2023. At the time of writing, this update was still in a release candidate stage, with an ongoing **request for comment** (**RFC**) in place via the Open Worldwide Application Security Project GitHub repository (`https://github.com/OWASP/API-Security/tree/master/editions/2023/en`).

Let us take a quick look at the currently proposed Top 10, shown in summary here:

#	2019	2023
API1	Broken Object Level Authorization	Broken Object Level Authorization
API2	Broken User Authentication	Broken Authentication
API3	Excessive Data Exposure	Broken Object Property Level Authorization
API4	Lack of Resources & Rate Limiting	Unrestricted Resource Consumption
API5	Broken Function Level Authorization	Broken Function Level Authorization
API6	Mass Assignment	Server Side Request Forgery
API7	Security Misconfiguration	Security Misconfiguration
API8	Injection	Lack of Protection from Automated Threats
API9	Improper Assets Management	Improper Inventory Management
API10	Insufficient Logging & Monitoring	Unsafe Consumption of APIs

Table 3.1 – Open Worldwide Application Security Project API Security Top 10 proposed changes

The changes are shown highlighted and the first observation is that there are several flaws and severities that remain unchanged, particularly at the top of the ranking. BOLA and broken (user) authentication remain the top severities impacting APIs.

Let us review the four new entries in the list for 2023:

- **API3:2023—Broken Object Property Level Authorization**: This new flaw type is not actually new; rather, it is a combination of two existing flaw categories, namely **Excessive Data Exposure** and **Mass Assignment**. This new category can be considered a flaw affecting the reading or writing of properties (typically data) within an API. Based on empirical data from my research, **Excessive Data Exposure** (data read) issues dominate **Mass Assignment** (data write) issues.

- **API6:2023—Server Side Request Forgery**: This may be a new category of flaw for the Open Worldwide Application Security Project API Security Top 10 but is well known to AppSec teams worldwide. This flaw occurs when an API backend accepts a client-supplied path or URL without validating it, thereby allowing an attacker to control the behavior of the API by deliberately supplying URLs under their control. This allows an attacker to exfiltrate data, overwrite data and configuration, or cause misoperation of the API itself.

- **API8:2023—Lack of Protection from Automated Threats**: This flaw relates to the abuse and misuse of APIs by automated agents, typically in the form of bots or distributed attacks via scripts or automation. Since the bot attempts to automate normal behavior, it can be difficult to distinguish an automated attack via bots from the normal use case. Also, automated attacks often do not require any API vulnerability to succeed. Instead, they use the API in the way it was intended but just at a very high volume or frequency. This new flaw category replaces the `API8:2019 Injection` flaw, which drops out of the Top 10 in 2023.

- **API10:2023**—Unsafe Consumption of APIs: The final addition to the top 10 list is an interesting one that will be familiar to AppSec professionals familiar with the broad category of trusting user (and unvalidated) input. The **Unsafe Consumption of APIs** flaw occurs when client API implementations consume upstream APIs without validating the responses, leaving themselves open to vulnerabilities inherited from the upstream API.

This very quick preview of the 2023 candidate release of the Open Worldwide Application Security Project API Security Top 10 indicates that API security is a rapidly changing landscape where some things stay the same while many things change.

Summary

It was a long journey through this chapter—by now, you will have a strong understanding of flaws, vulnerabilities, and threats and how they present a risk to your APIs. The key vulnerabilities that affect APIs include broken object-level and function-level authorization, broken authentication, data vulnerabilities, and finally, implementation and configuration vulnerabilities. Even if your API is free of vulnerabilities, it can still be abused or susceptible to business logic attacks.

While this chapter has focused on the theoretical nature of vulnerabilities, we are about to see just how easily these vulnerabilities can result in major breaches as we take a deep dive into twelve recent API breaches.

Further reading

Further reading on CWEs and the Open Worldwide Application Security Project is available here:

- `https://cwe.mitre.org/top25/archive/2022/2022_cwe_top25.html`
- `https://Owasp.org/www-project-api-security/`
- `https://owasp.org/www-pdf-archive/API_Security_Top_10_RC_-_Global_AppSec_AMS.pdf`
- `https://christianespinosa.com/blog/risk-comprehension-is-a-basic-cybersecurity-skill-yet-most-practitioners-lack-it/`

Further reading on abuse cases is available here:

- `https://blog.cloudflare.com/api-abuse-detection/`
- `https://theauthapi.com/articles/right-ways-of-rate-limiting/`
- `https://www.darkreading.com/dr-tech/even-perfect-apis-can-be-abused`
- `https://portswigger.net/web-security/logic-flaws/examples`

Investigating Recent Breaches

One of the best ways to learn is through the experience of others, particularly if these experiences come with deleterious consequences. Unfortunately, in the case of API security, there is no shortage of breaches and security incidents from which to learn.

In this chapter, we will look at a number of real-world API breaches, gaining an appreciation of how poor design decisions and implementation flaws allowed a skilled attacker to exploit the API, often with serious consequences. Explore this chapter with an open mind and consider, at each point, the decisions made and how you would have done things differently. Remember that hindsight is 20/20 – it's not always as easy as it might seem.

In this chapter, we will examine several recent real-world breaches, focusing on the following areas:

- What happened with the incident?
- What was the root cause of the incident?
- What was the impact of the incident?
- How can such incidents be prevented in the future?

The importance of learning from mistakes

Previously, in my career as an application security consultant in large organizations, I struggled to convey the importance of a particular security concept or principle to a development team. Although developers can appreciate the theory of an underlying issue (for example, SQL injection), there can be a perception that the issue is not a real-world example and would not occur in practice. The most powerful tactic in such cases is to provide a practical hands-on demonstration of the vulnerability, and then, even more impactful, is to demonstrate how this issue impacted an organization by showcasing details of a breach or incident.

We are living in an age of greater transparency in the disclosure of security incidents, thanks to the advent of responsible disclosure initiatives and managed bug bounty programs. Organizations are far more likely to disclose the nature of a breach and the actions they have taken to prevent a recurrence, rather than the tedious cycle of denial, obfuscation, and procrastination. Human nature is inclined to be forgiving in the face of honesty, particularly if coupled with an admission of error and a commitment to improving.

One of the core tenets of a mature DevOps organization is the practice of *The Third Way: Principles of Continuous Learning* (a link to read more on this is in the *Further reading* section at the end of this chapter). As organizations develop products on ever-shortening release cycles, the likelihood of new defects (such as security vulnerabilities) being introduced increases accordingly. To avoid a reduction in quality, organizations need to ensure that they are proactive in the following areas:

- Finding and fixing defects as early as possible through inspection and testing
- Learning lessons from mistakes to avoid a repetition of the same mistake

Blameless postmortems are a key tenet of the **site reliability engineering** (**SRE**) culture pioneered by Google's engineering teams. Such postmortems are an excellent opportunity for teams to learn from an incident and to incorporate practices focused on the prevention of future similar incidents. The key to conducting these postmortems is that they are blameless – they focus on the causes or issues rather than on the team or individual. Here is Google's take on this:

> *For a postmortem to be truly blameless, it must focus on identifying the contributing causes of the incident without indicting any individual or team for bad or inappropriate behavior.*

I am deliberately laboring the importance of focusing on the issue rather than the individual. If blame is apportioned to individuals, then they will avoid admitting or owning up to mistakes in the future. This denies the possibility of both individual and organizational learning.

In this spirit of seeking to understand the reasons and causes of API vulnerabilities, let's embark on our learning journey with an inquisitive mindset.

Examining 10 high-profile API breaches from 2022

In my professional role, I produce a weekly newsletter on API security topics at APISecurity.io (`https://apisecurity.io/`). From my work at APISecurity.io, I have picked ten of the top breaches from 2022, which give a representative sample of real-world API vulnerabilities and how they lead to the loss of data or personal information. Let's get started.

> **Errors and omissions excepted**
>
> Information in this section is taken from publicly disclosed sources, including bug reports, vulnerability tracking sites, first- and third-party blogs and research sites, and industry news websites.
>
> To the best of the my knowledge, the information is accurate at the time of writing; however, as is the nature of technology, the landscape changes rapidly, and new information or details may have come to light in the interim.

1–Global shipping company

In February 2022, security researchers at Pen Test Partners disclosed details of an API vulnerability on the website of a popular global shipping company.

What happened?

Researchers discovered they could use automation to submit random parcel numbers (the format of the parcel numbers conformed to a well-known standard pattern) to an API that retrieved a map image showing the approximate location of the recipient, as shown here:

Figure 4.1 – Location of recipient

In the United Kingdom, it is relatively easy to reverse engineer a postal code (or zip code) by using street names and approximate locations. The researchers discovered they could use the reverse-engineered postal codes and their guessed parcel number to impersonate the recipient to the API, which then returned a more complete history of the parcel delivery, as shown in the following screenshot:

Your parcel has been delivered and received by FRED at 12:10pm on Wed 8 Sep 2021

Click on image to view on map

Figure 4.2 – Full parcel tracking details

As any good researcher would do, they inspected the API traffic using the browser tools and the **JSON** viewer and revealed the following:

```
Body    Cookies    Headers (4)    Test Results

Pretty    Raw    Preview    Visualize    JSON  ∨

1    {
2        "addressPoint": {
3            "longitude": "-0.09267161261982491",
4            "latitude": "51.51146210820201"
5        },
6        "notificationDetails": {
7            "mobile": "555-555-5555",
8            "email": "fred@acme.com",
9            "contactName": "Fred Rogers"
10       },
11       "podDetails": {
12           "podName": "Tom",
13           "signatureRequired": true
14       }
15   }
```

Figure 4.3 – Full recipient details, including PII

The API returned full recipient details in addition to the package destination, including the phone number, email, and exact address – a rather significant disclosure of **personal identifiable information (PII)**.

What was the impact?

The security researchers immediately notified the shipping company, who investigated and released a fix a month later. The shipping company does not believe this vulnerability was exploited by any attacker, although it is possible this information was used in phishing and impersonation attacks, which were extremely common during the COVID-19 pandemic, particularly for shipping companies.

At the request of the shipping company, the security researchers delayed the publication of their report until after the busiest season (Christmas).

What was the root cause?

There are two main API vulnerabilities here:

- **Lack of rate-limiting**: The API used to render the map image did not incorporate any form of rate-limiting, which allowed attackers to use a brute force method to guess parcel numbers based on a pattern. Using rate-limiting would have significantly slowed down the attacker's progress.

- **Excessive information exposure**: The more serious issue was with the second API, which returned the detailed tracking information. For the purposes of the website, only the estimated delivery time, current location, driver name, and rating were needed. Unfortunately, the API returned a full record of the shipment, including PII, which was certainly not needed.

There are two other anti-patterns on display here:

- Never rely on the client side or frontend to suppress the display of unnecessary or superfluous data. Any data transmitted to the client is easily examinable using even basic tools built into the client – in this case, the browser tools.

- The developers failed to understand the threat landscape adequately – although an approximate map image may seem benign, it was relatively easy to weaponize this against the API.

How could this have been prevented?

Firstly, API developers are advised to use rate-limiting on all sensitive API endpoints. Rate-limiting is a robust technology that can be implemented in the API code, in an API gateway, or in an API firewall. The rate-limiting algorithm can limit based on a sliding window or a progressively longer back-off timeout to ensure normal operations are not impacted. Now, which API endpoints are sensitive? Anything that will accept user input (particularly if that input can be easily guessed) is vulnerable to brute force abuse. Typical endpoints include password reset or registration functions.

Secondly, in the API design phase, ensure that data privacy considerations are enforced. In this example, the API was excessively verbose – a design review would have easily shown that PII data was being returned unnecessarily. The key is to understand the sensitivity of the data and then to understand the intended recipient. If the recipient should not have access to sensitive data, then the API must be refactored to return the minimum data to which they have access. Developers often feel they are being helpful by returning more data than needed and fail to realize the negative impact of this practice. Technologies such as API firewalls and gateways can be used to enforce a data contract according to the **OpenAPI Specification** (**OAS**) definition, to prevent excessive leakage.

Also, never, ever rely on client-side controls to mask or hide data – it can always be bypassed using simple browser tools.

Further reading on this vulnerability

The Bleeping Computer website has provided an excellent write-up of the reported vulnerability, and you can read it for full details. The article is available here:

```
https://www.bleepingcomputer.com/news/security/dpd-group-parcel-
tracking-flaw-may-have-exposed-customer-data/
```

2–Campus access control

The second vulnerability, from March 2022, affected the API of a campus access control system, which allowed an attacker to create an electronic **master key**.

What happened?

A student at a US college campus had become frustrated with the poor performance of the application used to control his room access on campus. To overcome the application limitations, he applied simple reverse engineering techniques to understand the backend API.

The student quickly discovered the relevant API method, which required the caller to submit a precise location to ensure that a door could only be unlocked if the caller was in close proximity to it. The student then explored additional APIs, all of which took a student ID as a parameter. Unfortunately, these IDs are easily guessable or available in campus directories, so are hardly a secret value.

The student then made the remarkable discovery that endpoints taking student IDs did not require any further form of authentication, simply the ID was sufficient to allow API access. Armed with this discovery and the fact that a location could easily be faked, the student was able to demonstrate a viable master key for any lock on any campus using the same software.

What was the impact?

The student initially attempted to contact the vendor of the application; however, after some months, he received no response. He then contacted TechCrunch (`https://techcrunch.com`) to ask them to share the details with the vendor. The vendor then released an update to fix the issue and revoked tokens. The vendor advised that affected customers had been notified but declined to provide further details.

At the time of writing, it was unknown whether the vulnerability had affected any installations as the vendor refused to respond to requests for access logs.

What was the root cause?

This is a textbook example of broken authentication, which is one of the highest-impact API vulnerabilities. In this case, it is actually worse than simply broken authentication; there was no authentication in place at all.

How could this have been prevented?

A vulnerability of this nature is absolutely avoidable – all API endpoints other than public APIs must have adequate authentication. Standard mechanisms for authentication exist (such as **OAuth2**), or even API keys or tokens provide significant protection. In this case, even though a basic username/password login was required by design, the implementation failed to validate the password.

Code inspections, linting of API endpoints, and OAS definition audits can help reveal flawed or missing authentication in endpoints.

Further reading on this vulnerability

The TechCrunch website has provided an excellent write-up of the reported vulnerability, and you can read it for full details. The article is available here:

`https://techcrunch.com/2022/03/03/cbord-university-digital-locks/`

3–Microbrewery application

The next vulnerability comes courtesy of Pen Test Partners and affected the mobile application and backend API of a popular UK microbrewery, allowing the disclosure of PII data.

What happened?

While reviewing the mobile application for the microbrewery, security researchers discovered that the developers had embedded bearer authentication tokens directly into the binary of the application. By using simple inspection, they discovered these three tokens:

```
getUser:function(t){return
o.default.get("https://www.brewdog.com/uk/rest/uk/V1/customers/"+t,{headers:{'Cache-
Control':'no-cache, no-store, must-revalidate',Pragma:'no-cache',Expires:0,Authorization:"bearer
y99a5p6dhqspwr51h5z9r6h7t0zuaw5x"}})},

getUserWithUsername:function(t){return
o.default.get("https://www.brewdog.com/uk/rest/uk/V1/customers/search?searchCriteria[filterGro
ups][0][filters][0][field]=email&searchCriteria[filterGroups][0][filters][0][value]="+t+"&searchCriteria
[filterGroups][0][filters][0][conditionType]=equals",{headers:{'Cache-Control':'no-cache, no-store,
must-revalidate',Pragma:'no-cache',Expires:0,Authorization:"bearer
y99a5p6dhqspwr51h5z9r6h7t0zuaw5x"}})},

setMyLocal:function(t,s,n){return
o.default.put("https://www.brewdog.com/uk/rest/uk/V1/customers/"+t.id,{customer:{id:t.id,group
_id:t.group_id,email:t.email,firstname:t.firstname,lastname:t.lastname,store_id:t.store_id,website_i
d:t.website_id,custom_attributes:[{attribute_code:'my_local_id',value:s},{attribute_code:'my_local_
reset_date',value:n}]}},{headers:{Authorization:"bearer y99a5p6dhqspwr51h5z9r6h7t0zuaw5x"}})}};
```

Figure 4.4 – Embedded hardcoded tokens

Ordinarily, such bearer tokens would be obtained via a token exchange protocol such as OAuth 2. Even if tokens are stored locally, they should at least be encrypted to prevent them from being easily viewed.

The researchers realized that if the tokens were hardcoded in the application, then all application users shared the same token. The only thing that differentiated users was their customer ID. In the case of this application, the IDs were easily manipulated via incrementing through a range, allowing the researchers to access the accounts of other users. They also discovered that the API was exposing verbose information, including PII data, which could lead to GDPR violations if leaked to unintended recipients. Here is an example of a typical extract:

Figure 4.5 – PII data extract from API

What was the impact?

It is possible that over 200,000 stockholders of "Equity for Punks" have had their personal information exposed as a result of this vulnerability. Currently, it is unclear whether the vulnerability has been exploited, but the potential impact is considerable. Additionally, an attacker could have forged discount codes exceeding their correct value or used the birthday discount scheme to get free beer from the brewery.

What was the root cause?

The root cause was the use of hardcoded tokens stored in the clear in the application binary. This is probably one of the most egregious instances of poor software security featured in the newsletter. A secondary issue was the excessive exposure of customer data via the API.

How could this have been prevented?

There are several recommendations to prevent a vulnerability like this:

- Application developers should use a robust, industry-proven framework, such as OAuth 2, rather than an inadequate method, such as hardcoding the credentials.

- Application and API code should be scanned with tools (such as **grep**) to identify hardcoded credentials within a code base. This is a low-cost solution that can be easily integrated into the **CI/CD** process.

- Because the researcher was able to manipulate the customer IDs, there is a likelihood that the API may have been vulnerable to **broken object-level authorization** (**BOLA**). The application developer needs to ensure that all API requests are fully authenticated and authorized.

Further reading on this vulnerability

The discoverers of this vulnerability, Pen Test Partners, have provided an excellent write-up of the reported vulnerability, and you are advised to consult it for full details. The article is available here:

```
https://www.pentestpartners.com/security-blog/free-brewdog-beer-
with-a-side-order-of-shareholder-pii/.
```

4–Cryptocurrency portal

The most significant API vulnerability we cover in this book is one that affected a cryptocurrency portal, which allowed the potential of unlimited currency transactions without backing assets – a license to print money!

What happened?

A researcher discovered that he could exploit a vulnerability in the portal by using two different cryptocurrency accounts with a modest balance in one account, and zero in the other. He could initiate a market order using the account containing funds as the source account but could then modify the API request to specify the other account with a zero balance. Unfortunately, the validation logic did not verify the source account properly and processed the trade normally. Thus, the researcher could complete the trade using cryptocurrency they did not have.

The researcher contacted the platform via X, and they responded immediately. Within six hours, the researcher and platform security team resolved and verified the vulnerability.

What was the impact?

The impact was fortunately very minimal due to the rapid and responsible disclosure of the vulnerability. No loss is thought to have been incurred.

The platform did issue additional guidance stating that they had other compensating controls in place to limit high-value transactions via trade circuit breakers and other continuous anomaly detection methods.

Bug bounties can be profitable – this was worth $250,000 to the researcher.

What was the root cause?

This vulnerability is one of the best cases of broken object-level authorization. Although the API endpoint processing the transaction was authenticating the user correctly, the endpoint failed to check whether the requested source account had sufficient assets for the trade. This allowed the researcher to use one account as the source of the validation for sufficient assets but another account as the actual source of the trade.

How could this have been prevented?

Broken object-level authorization vulnerabilities must be addressed at the API backend implementation layer. The backend should not trust the integrity of submitted parameters (such as an account ID) since these could be manipulated. Always fully validate that the authenticated user has the requisite access to the requested object.

This topic is covered in much more detail in *Part 3, Defending APIs*.

Further reading on this vulnerability

The affected vendor has provided a disclosure of the reported vulnerability, and you are advised to consult this for full details. The article is available here:

```
https://blog.coinbase.com/retrospective-recent-coinbase-bug-bounty-
award-9f127e04f060
```

5–Dating application

One of the best technical (and most entertaining) write-ups came from a security researcher who discovered several API vulnerabilities in two different dating applications: Tinder and Bumble.

What happened?

The researcher began his investigation by attempting to exploit the user location features in the Tinder application. Frequently, mobile applications using location are vulnerable to attacks leading to leakage of the user's precise location.

The researcher first examined location information available via the API and discovered that locations with extremely high precision were being used, as shown in this example:

```
{
   «user_id»: 1234567890,
   «location»: {
      «latitude»: 37.774904,
      «longitude»: 122.419422
   }
// ...etc...
}
```

Using relatively simple trigonometry, it is possible to use three locations to perform **trilateration** on a user to determine their location. In a case where high-precision coordinates are returned, it is possible to establish a precise user location. This diagram shows the principle of trilateration:

Figure 4.6 – Using trilateration to calculate user location

In order to perform the calculation, the researcher needed three sets of coordinates. Fortunately, he found another bug in the API implementation, which leaked user IDs within the application UI. Using these IDs, he was able to retrieve location information without even being connected to that user.

The researcher also discovered a weakness in the application's hashing functions, which were used to protect data in transit. The application developers had made the mistake of implementing the hashing function within JavaScript bytecode on the client, which could easily be reverse-engineered.

What was the impact?

This vulnerability was not thought to have impacted any users, although the disclosure caused embarrassment for the application affected.

What was the root cause?

The primary vulnerability, in this case, was broken object-level authorization. By solely knowing the user ID, the researcher was able to retrieve their full profile and location information.

A second issue was related to excessive information exposure via the API. In this case, it was exposing data of unnecessary precision, which could allow a second-order attack such as trilateration.

The application also relied on **security by obscurity**, by attempting to mask implementation details on the client side. Any skilled attacker can easily reverse-engineer such techniques.

How could this have been prevented?

There are several recommendations to prevent a vulnerability like this:

- This attack could have been thwarted by detecting an excessive number of potentially impossible movements for a single user account. The attack relied on being able to spoof user locations at a fairly rapid rate, and this could have been detected by a rate-limiting algorithm on the API.

- As per other vulnerabilities described earlier, never rely on client-side controls as these can be bypassed or reverse-engineered. Also, avoid relying on security by obscurity for the same reasons.

- In this instance, we had a subtle variation on the theme of excessive information exposure. Rather than returning unnecessary data, the API returned information with excessive precision. These types of issues can easily be detected at design time and enforced within the OAS definition.

Further reading on this vulnerability

The researcher who found this vulnerability has provided an excellent (and very humorous) write-up of the reported vulnerability, and you can read it for full details. The article is available here:

```
https://robertheaton.com/bumble-vulnerability/
```

6–The All in One SEO WordPress plugin

The popular CMS platform WordPress is frequently in the news due to security vulnerabilities, often caused by insecure plugins.

What happened?

A popular WordPress plugin called All in One SEO was discovered to be vulnerable to an authenticated privilege escalation attack.

The WordPress architecture allows plugins to be activated at runtime to extend the platform's functionality. A plugin must implement various standard methods to validate API requests and, in this case, the `validateAccess()` method contained a vulnerability, as seen in the following:

```
219 /**
220  * Validates access from the routes array.
221  *
222  * @since 4.0.0
223  *
224  * @param  \WP_REST_Request $request The REST Request.
225  * @return bool                      True if validated, false if not.
226  */
227 public function validateAccess( $request ) {
228     $route     = str_replace( '/' . $this->namespace . '/', '', $request->get_route() );
229     $routeData = isset( $this->getRoutes()[ $request->get_method() ][ $route ] ) ? $this->get
230 
231     // No direct route name, let's try the regexes.
232     if ( empty( $routeData ) ) {
233         foreach ( $this->getRoutes()[ $request->get_method() ] as $routeRegex => $routeInfo )
234             $routeRegex = str_replace( '@', '\@', $routeRegex );
235             if ( preg_match( "@{$routeRegex}@", $route ) ) {
236                 $routeData = $routeInfo;
237                 break;
238             }
239         }
240     }
241 
242     if ( empty( $routeData['access'] ) ) {
243         return true;
244     }
245 
246     // We validate with any of the access options.
247     if ( ! is_array( $routeData['access'] ) ) {
248         $routeData['access'] = [ $routeData['access'] ];
249     }
250     foreach ( $routeData['access'] as $access ) {
251         if ( current_user_can( $access ) ) {
252             return true;
253         }
254     }
255 
256     if ( current_user_can( apply_filters( 'aioseo_manage_seo', 'aioseo_manage_seo' ) ) ) {
257         return true;
258     }
259 
260     return false;
261 }
```

Figure 4.7 – All In One plugin validator

The vulnerability occurs on line 235, where a case-insensitive path string is used. By changing a character case in the requested URL, the `validate` method would execute line `243` of the validation function and return a `true` value, allowing full, privileged access to the WordPress installation. Once the attacker has gained privileged access, they could launch other attackers, such as SQL injection.

What was the impact?

The impact was potentially severe, reflected in the **Common Vulnerability Scoring System** (**CVSS**) score of 9.9 awarded to this vulnerability. Fortunately, the researchers who disclosed the vulnerability issued a bulletin to warn customers and released a new, patched version within a week. No impact on users was reported at the time.

What was the root cause?

This is a good example of broken function-level authorization and poor default settings in the API endpoint handler. By manipulating an input parameter (by changing the case of a character), an attacker could bypass authorization.

How could this have been prevented?

Great care should be taken while implementing critical security-sensitive endpoints such as those validating access. Sensible defaults should be used, namely denying access by default. In this example, a failure to validate reached an error handler that returned an `allow` value rather than a `deny` value.

Further reading on this vulnerability

The popular WordPress provider Jetpack has provided a write-up of the reported vulnerability, and you can consult it for full details. The article is available here:

```
https://jetpack.com/blog/severe-vulnerabilities-fixed-in-all-in-one-
seo-plugin-version-4-1-5-3/
```

7–X account information leakage

Throughout 2022, there were various reports indicating that the X social media platform had been subject to the loss of customer data, including usernames, email addresses, and telephone numbers. Initial estimates of the volume of data lost were numbers ranging from a few million, with the most recent estimates nearer to the 200 million mark.

What happened?

The leak originated via an internal API used for mobile device logins. The X backend exposes a number of so-called login **flows** to allow a mobile device to query the backend for account and user information. A typical example is a check to see if the username already exists when trying to sign up to X.

A researcher discovered that the login flow could be abused to test for the existence of arbitrary accounts using a `suggestion_id` to check against a provided email or telephone number. In the example shown, the attacker provides a `flow_token` in step 1 and a query string in step 2. If the identity in the query string exists, then a response code is returned, which includes the `user_id` of the account in step 3.

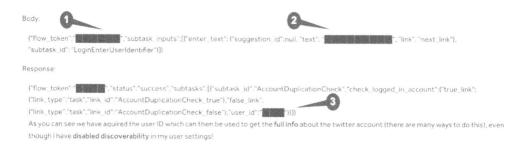

Figure 4.8 – X login flow sample

Although at first glance this issue seems totally benign, consider that an attacker using a massively scaled attack could build a mapping of all publicly known emails and/or telephone numbers against their associated X handle. This is very valuable information in the hands of an attacker operating a spear phishing campaign.

What was the impact?

At the time of writing, it is unclear whether X or its users have been victims of any phishing campaigns. Certainly, reports online indicate that up to 200 million records were available for sale online. X initially denied the reports and then admitted that account information leakage had occurred. Since no credentials or tokens were leaked, an attacker would need to launch a secondary attack to leverage these account details, thereby minimizing the overall impact.

What was the root cause?

This leakage was the result of an attacker discovering an undocumented API and then abusing that to achieve mass data exfiltration via the API. There is no actual flaw in the API implementation (other than that this API should have been disabled if the user had selected to disable discoverability in their privacy options); certainly, there are none of the **OWASP API Security Top 10** issues. The root cause was that the API designers failed to anticipate an abuse case where their API could be used in an unintended way to achieve a considerable amount of exfiltration.

How could this have been prevented?

There was no flaw within the implementation; it behaved exactly as designed. However, the designers overlooked the fact that a clever attacker could abuse this API to get high volumes of data, which would be attractive on the black market. If the designers had considered this outcome, they could have removed this function or at least ensured the sensitive user_id field was removed from the response.

Inevitably, APIs will be abused and misused, and it is increasingly important that security monitoring teams are able to detect such use cases and respond accordingly. In this particular example, the API endpoint would have been generating high volumes of 404: Not Found errors, which could have been easily detected by an API monitoring platform.

Further reading on this vulnerability

The popular bug bounty platform HackerOne has the full details of the reported vulnerability, and you can consult it for full details. The article is available here:

```
https://hackerone.com/reports/1439026
```

8–Home router

Another embedded system vulnerability is up next; this time in a home router.

What happened?

A popular home router was vulnerable to a command injection attack in an internal API. A security researcher discovered an internal admin interface that the router UI used to execute a `ping` command and was able to execute arbitrary commands by injecting additional characters after the target IP address.

The vulnerability received a CVSS score of 9.8, and at the time of writing, it was unknown whether the issue had been patched.

Home routers are often built on Linux-based operating systems such as **BusyBox**. Commonly, they expose additional administrative or diagnostic capabilities, such as the ability to ping IP addresses or run speed tests. In this case, the user interface offered the ability to ping an IP address, as shown in this example:

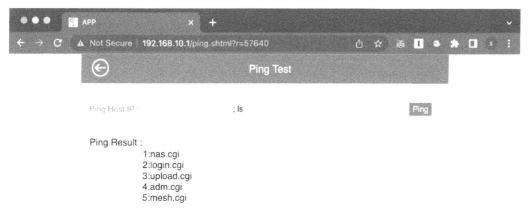

Figure 4.9 – Router ping interface

The researcher found that he was able to inject arbitrary commands via this interface (in this example, the command executed was `ping ; ls`, which shows a directory listing on the router).

The user then explored the router API using **Burp Suite** (shown in *Figure 4.10*) and discovered the /cgi-bin/adm.cgi endpoint, which allowed the execution of OS commands on the router. To his surprise, the researcher discovered that even if he removed the session cookie, he was still able to execute commands. He had discovered an unauthenticated endpoint, as depicted in the following:

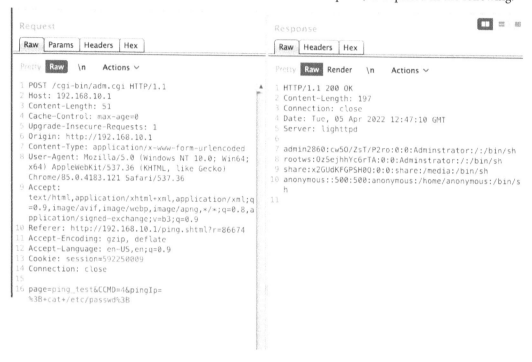

Figure 4.10 – Command API interface on the router

Additionally, the router did not enforce any **cross-site request forgery** (**CSRF**) protection, making the router vulnerable to attacks beyond the local network.

What was the impact?

No reported breaches were disclosed, and at the time of writing, no patches were available.

What was the root cause?

This is a case of broken authentication, where an exposed API endpoint allowed access without authentication.

Although an unauthenticated API endpoint is always of concern, what raised this to a high severity was the fact that the endpoint was also susceptible to command injection. This allowed an attacker to leverage the unauthenticated endpoint to inject arbitrary commands at the **root** access level.

How could this have been prevented?

The vulnerabilities present on this router are typical of the product category. Often, these products are produced on very tight budgets and inherit default OS images from chip vendors. Technical debt is inherited unless the vendor pays specific attention to security.

Shipping an unauthenticated API endpoint in a product is entirely preventable. Even the most basic of authenticated testing would have revealed this flaw.

Using recommended protections for CSRF would have prevented this exploit from being accessible through the public internet. Good defensive coding provides in-depth defense, which can thwart complex multi-stage attacks.

Further reading on this vulnerability

The researcher has provided a write-up of the reported vulnerability, and you can consult it for full details. The article is available here:

```
https://stigward.medium.com/wavlink-command-injection-cve-2022-
23900-51988f6f15df
```

9–Remote access to two popular vehicles

The penultimate breach is the result of research from Sam Curry regarding an API vulnerability affecting two popular vehicle ranges. The researcher created a basic proof of concept that demonstrated how the vehicle engine could be controlled remotely, including controlling the engine and vehicle door locks.

What happened?

Curry discovered he could access vehicle computers on two popular brands of vehicles and execute remote commands such as sounding the horn or starting the vehicle by manipulating the login to the web-based management platform.

Two main issues were at play: a lack of input sanitization and poorly constructed email validation regular expressions. Curry found that once he had a valid registration on the vehicle's web portal, he could modify the registration email address (by adding control characters), which would issue a **JSON Web Token** (**JWT**) without proper validation using this "hacked" email. By knowing or guessing a user's email, an attacker could successfully gain a token, allowing full vehicle access.

The following two screenshots demonstrate the manipulation of an email address and how this can be used to gain a valid JWT for vehicle access. The first shows the request for the token:

Figure 4.11 – Manipulation of email address in token creation

The request results in a token being issued, as seen in the following:

Figure 4.12 – Gaining vehicle access with a forged JWT

What was the impact?

Given the potentially catastrophic impact of a malicious actor gaining vehicle access using this vulnerability, it was fortunate that the researcher disclosed the vulnerability to the manufacturers in a responsible manner. The consequences of such vulnerabilities being available in the wild are too horrific to contemplate — perhaps, as one commentator remarked, "Disconnect your vehicle from the internet."

What was the root cause?

There are two fundamental issues at play with this implementation:

- Never, ever trust user input: In this case, an attacker provided a malformed version of an email address that was in use on the platform and trusted by the backend API

- Only ever use hardened regular expressions: The email validation methods accepted email addresses that were obviously invalid

How could this have been prevented?

Firstly, the API developers could have applied basic application security hygiene (such as using hardened regular expressions and testing corner cases on malformed data) and established best practices to eliminate the vulnerabilities observed.

Connected vehicle manufacturers should consider how they allow access to critical vehicle control systems. In the instance of high-privilege actions such as engine control, an out-of-band control mechanism might make sense.

Further reading on this vulnerability

The popular security news website Daily Swig has provided a write-up of the reported vulnerability, and you can consult it for full details. The article is available here:

```
https://portswigger.net/daily-swig/critical-vulnerability-allowed-
attackers-to-remotely-unlock-control-hyundai-genesis-vehicles
```

10–Smart Scale

Congratulations if you have made it all the way through the last nine vulnerabilities! We have kept the best for last – one of the most popular features in the APISecurity.io newsletter is the Yunmai Smart Scale disclosure.

What happened?

A penetration testing company in London conducted assessments on the Yunmai Smart Scale Android and iOS applications. In doing so, they discovered five separate API vulnerabilities:

- Bypass the limit of 16 family members per primary account

- User ID enumeration

- Ineffective authorization checks

- Information leaks

- Account takeover through password reset functionality

By combining three of these vulnerabilities, they were able to achieve account takeover on the platform.

The second vulnerability of the five identified allowed arbitrary enumeration of user IDs by guessing IDs. APIs were not sufficiently authorized to access the guessed ID and instead returned full user information, which included sensitive PII. Additionally, it returned the parent ID, which could be used in subsequent attacks. This enumeration can easily be automated using Burp Suite, as shown in the following screenshot:

Figure 4.13 – Enumeration of user IDs

The third vulnerability is a great example of broken authentication, which allowed the researchers to add and delete users from other people's accounts. In this case, all the researchers had to do was enumerate user IDs by guessing them. This is shown in the following screenshot, where an attacker can submit an arbitrary user ID to be deleted, despite not having the requisite permission for that resource:

Figure 4.14 – Deletion of user accounts

The researchers also discovered that when they created child accounts, the API leaked significant privileged information such as access tokens and, more critically, refresh tokens. Leaking an access token is a significant issue since an attacker can use that token for the duration of its validity to attack the associated target. However, leaking a refresh token is almost unforgivable from a security perspective, since this grants an attacker the ability to regenerate access tokens in perpetuity. The only way that the system can be protected in this case is to invalidate all existing sessions, including refresh tokens, which will then force clients to re-authenticate. The leakage of the refresh token is highlighted in the following screenshot:

Figure 4.15 – Excessive information exposure leaking tokens

The researchers also discovered they could abuse the **forgotten password** feature of the API by submitting a sequence of viable reset PINs (unfortunately, the reset PINs had insufficient entropy and were quite predictable). The backend API failed to block attempts to brute force this critical API endpoint. This is shown in the following screenshot:

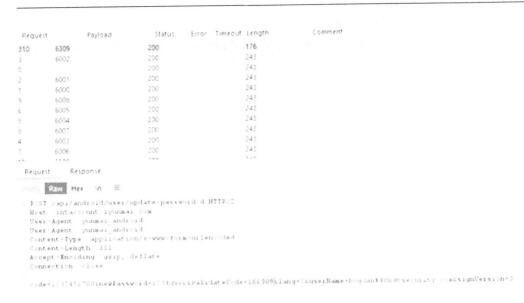

Figure 4.16 – Brute forcing of password reset functionality

What was the impact?

No end users were affected since the researchers disclosed their findings to the vendor. It took the vendor several attempts to remediate the findings, and after several re-tests, only some of the issues had been addressed. At the time of writing, the user ID enumeration and ineffective authorization checks were still unresolved. The researchers' findings were then disclosed publicly.

What was the root cause?

This write-up is a great demonstration of how multiple API vulnerabilities can be chained together to achieve far greater impact than any one individually. This is a core tenet of software security – individual flaws may seem benign and not worthy of remediation; however, in the hands of a skillful adversary, they can be combined to achieve catastrophic consequences.

In summary, these were the three OWASP API Security Top 10 vulnerabilities:

- Broken object-level authorization allowed the researchers to spoof user IDs and access accounts not belonging to the session user

- Broken authentication allowed the researchers to delete accounts and users that did not belong to them

- Excessive data exposure disclosed both access tokens and refresh tokens, and account and parent account IDs

How could this have been prevented?

Due to a large number of relatively basic API vulnerabilities, it appears the vendor's developer teams would benefit from API security training and awareness courses. Many of the issues disclosed are easy to remediate or avoid in the first instance.

Similarly, many of the vulnerabilities could have been discovered via automated API-specific testing before being released to production.

Further reading on this vulnerability

The researchers who discovered this set of vulnerabilities have provided an excellent write-up of the reported vulnerability, and you can consult it for full details. The article is available here:

`https://www.fortbridge.co.uk/research/mass-account-takeover-yunmai/`

Key takeaways and learning

Hopefully, by now, you understand how easily API vulnerabilities can be exploited. Although none of the vulnerabilities featured resulted in dire consequences or financial loss, this is in some cases down to good fortune.

Firstly, the root cause of nearly all the flaws in the APIs resulted from either human error (where developers made basic mistakes) or a lack of security skills (where developers lacked an understanding of how their APIs could be attacked). Both can be addressed by developer training in the form of computer-based lessons or instructor-led courses.

Secondly, many of the vulnerabilities could have been avoided in the first place using a combination of secure design (threat modeling and risk assessments), secure coding best practices, and, most importantly, testing at every stage of the software development life cycle.

Thirdly, most of the attacks used simple methods and were unsophisticated in nature. Certainly, no advanced tools were necessary, and with a little bit of dedication and perseverance, any tech-savvy user would have been able to perform these exploits themselves. This is a stark warning to API developers and defenders – the barrier to entry for your adversaries is low.

Finally, many of the most egregious flaws could have been detected using simple tools and techniques. For example, hardcoded tokens can be easily detected in code before committing to a repository, and unauthenticated endpoints can easily be discovered.

Summary

We have covered a lot in this chapter. By now, you should have a good understanding of how APIs can be attacked and how sometimes benign flaws can lead to severe compromise. On a note of caution, we have seen how easily flaws can be introduced and, likewise, how easily an attacker can use these flaws to compromise a system. On a more optimistic note, many of the flaws covered can easily be detected early in the development life cycle and, with education and sound design principles, can be avoided entirely.

With our sound grounding in the fundamentals of APIs, let's focus on how APIs can be attacked in the next section.

Further reading

Further reading on DevOps principles is available here:

- `https://postmortems.pagerduty.com/culture/blameless/`
- `https://blog.sonatype.com/principle-based-devops-frameworks-three-ways`

Further reading on API security is available here:

- `https://apisecurity.io/`
- `https://apisecurity.io/encyclopedia/content/owasp/owasp-api-security-top-10.htm`

Part 2: Attacking APIs

In this part, you will explore the techniques and approaches used by adversaries to attack APIs. This part is important for API defenders to understand how their APIs will be attacked, enabling them to defend themselves against common attack methods. The part covers the foundations of attacking APIs (covering common tools and methods), how to discover APIs in the real world, and then how to apply attack methods to overcome common API weaknesses and vulnerabilities.

This part has the following chapters:

- *Chapter 5, Foundations of Attacking APIs*
- *Chapter 6, Discovering APIs*
- *Chapter 7, Attacking APIs*

Foundations of Attacking APIs

In this chapter, the focus turns to the foundational issues associated with attacking APIs. Firstly, we will understand the different ways that an attacker can exploit an API using methods that include passive monitoring (discovery) and active interception, including the modification of requests and responses. We will then focus on a selection of the most important tools available to an aspiring API attacker and demonstrate how these can be used to perform core attacks, such as cracking passwords or tokens. Finally, we will combine this knowledge to build our own hacking laboratory and commence attacking some popular vulnerable APIs.

This chapter will equip you with the foundational knowledge used by API hackers—there are a vast array of tools and techniques available to a budding attacker, and it is important to know the relative value of different tools and techniques for a given attack scenario.

In a nutshell, this chapter is going to cover the following main topics:

- API attackers and their methods
- Mastering the tools of the trade
- Learning the key skills of API attacking

Technical requirements

For this chapter, you will need a development machine capable of the following:

- Running Docker locally
- Running VS Code with various marketplace extensions
- Access to the internet and a GitHub account to access the examples

This chapter contains many code samples in various languages; these can either be run locally, which will require the installation of compilers, SDKs, and frameworks, or from within a Docker build container.

The example code and various breaking changes to the instructions can be found in the `Chapter 5` folder on the book's GitHub repository here: `https://github.com/PacktPublishing/Defending-APIs/tree/main/Chapter5`

Understanding API attackers and their methods

In this opening section, we will examine the different methods a would-be attacker can employ to exploit an API, including passive and active traffic interception; finding API keys; fuzzing APIs for endpoints, their methods, and associated passwords; and cracking **JSON Web Tokens (JWTs)**.

Using an appropriate combination of these methods will allow an attacker to launch a successful discovery phase on an API from where further specific attacks can be launched.

Interacting with APIs

APIs are, by their nature, *headless*; in other words, they do not expose a user interface that can be used to exercise their functionality. An attacker has many methods to interact with an API to discover and exploit weaknesses. Usually, the first step will be to passively examine API traffic using an intercepting proxy (such as Burp Suite) or with an API testing tool such as Postman. The attacker's goal at this stage is to understand how the API works, the sequence of calls, the payloads, the authentication methods, and so on. *Figure 5.1* shows a simple setup where an API is exercised by a web or mobile application via an intercepting proxy—an attacker can observe both the requests and responses.

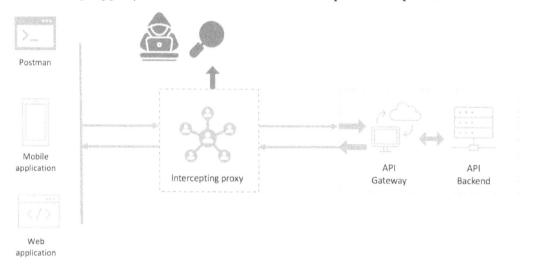

Figure 5.1 – Passive monitoring of API traffic

Once the attacker has an idea of how an API works, they can start actively intercepting the traffic using advanced features of the intercepting proxy. For example, in Burp Suite the attacker can intercept requests, modify parameters and requests, and submit them to the API. This process can be fully automated with random payloads allowing an attacker to rapidly execute a high volume of attacks. *Figure 5.2* shows the setup for the active interception of an API.

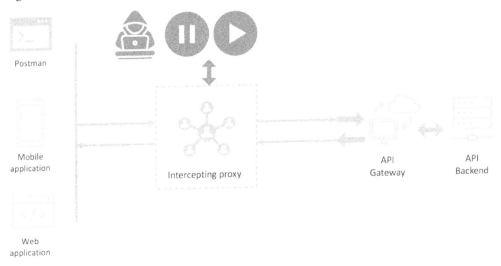

Figure 5.2 – Active interception and manipulation of API traffic

Finally, once an attacker has mapped out the API, they can directly access the API using a number of test tools that can perform advanced attacks, such as the brute-forcing of passwords, the discovery of API endpoints, the fuzzing of payloads, and so on. This is shown in *Figure 5.3*:

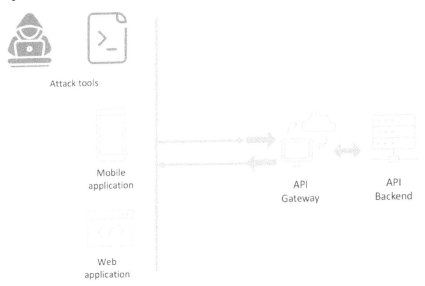

Figure 5.3 – Direct manipulation of API with tools and scripts

In reality, an experienced API attacker will use a combination of these methods depending on their goals.

> **Performing "machine-in-the-middle" attacks**
>
> While an intercepting proxy allows an attacker visibility into the API request and response traffic, an API and its client can be protected from this so-called **machine-in-the-middle (MitM)** attack method by using **Transport Layer Security (TLS)**. When a client makes a connection to an API, it will perform a TLS connection setup, which involves the exchange and verification of public key certificates. If an intercepting proxy is between the client and the API server, the certificate verification will fail since the certificate presented by the proxy will be self-signed and not trusted by default. To overcome this limitation, the attacker can install and trust the proxy's certificate on the client device, in which case the exchange will be intercepted by the proxy. Obviously, in a real-world situation, an attacker will not have access to the client device (hopefully); however, please bear this important step in mind when setting up your own intercepting proxy. This is often a source of frustration for learners. A client can further harden the security of the transport layer using a technique called *certificate pinning*, which binds the certificate to the domain name. This prevents the insertion of intermediate certificates.

Finding API keys

One of the first items an API attacker looks for is API keys or tokens that can be used to access authenticated API endpoints. If you can use an intercepting proxy (in other words, you have access to the network traffic and can install a trusted certificate on the client device if TLS is used), then you can trivially extract keys and tokens from the proxy.

Keys and tokens can be discovered in a wide range of locations, such as the following:

- Open S3 buckets on AWS
- Public GitHub repositories and *gists*
- Pastebin *pastes* (`https://pastebin.com/`)
- SharePoint, Confluence, and Jira sites
- Log files
- CI/CD systems (as unprotected variables)
- API request URLs

Once you have discovered a key or token, you should verify it is still valid by querying it against the relevant service endpoint. For example, to test a harvested Facebook token, use the following URL and check the status code:

```
https://developers.facebook.com/tools/debug/accesstoken/?access_
token=ACCESS_TOKEN_HERE&version=v3.2
```

One of the most popular tools for retrieving keys and tokens from GitHub repositories is TruffleHog (`https://trufflesecurity.com/blog/introducing-trufflehog-v3/`). The current version has a wide range of so-called detectors that can interpret credential types and then validate them against the appropriate endpoint. For example, the following code snippet showcases TruffleHog running as a container detecting keys in a GitHub test repository—in this case, a single AWS key is detected at the given URL:

```
colind@mbm1pro:~# docker run --platform linux/arm64 -it -v "$PWD:/pwd"
trufflesecurity/trufflehog:latest github --repo https://github.com/
trufflesecurity/test_keys
TruffleHog. Unearth your secrets.
Found verified result
Detector Type: AWS
Raw result: AKIAYVP4CIPPERUVIFXG
Link: https://github.com/trufflesecurity/test_keys/blob/
fbc14303ffbf8fb1c2c1914e8dda7d0121633aca/keys
Repository: https://github.com/trufflesecurity/test_keys.git
Commit: fbc14303ffbf8fb1c2c1914e8dda7d0121633aca
Email: counter counter@counters-MacBook-Air.local
File: keys
Timestamp: 2022-06-16 10:17:40 -0700 -0700
Line: 4
```

Once you have verified your harvested keys or tokens, you can use them to exploit the API—a good first step is to generate a fresh set of credentials in case your harvested ones are revoked.

Enumeration and discovery of APIs

When attacking an API, the first challenge is to determine the endpoints to be able to run an attack. One of the benefits of a REST API is that the designer can name endpoints to match their underlying data and entities, meaning there is no convention for their endpoints. How do you attack an API if you do not know the endpoints?

Fortunately, several excellent tools can perform endpoint discovery and effectively map out the API attack surface. My favorite of these tools is the excellent Kiterunner (`https://github.com/assetnote/kiterunner`), which runs a scan against an API to identify their existence and status code. Kiterunner uses proprietary databases built from a very large dataset of OpenAPI definitions harvested from the internet (primarily GitHub). While the databases cannot be guaranteed to be exhaustive, in practice, popular API frameworks (such as Express on Node.js and Flask in Python) use a fairly predictable set of endpoint names—by harvesting a large enough volume of OpenAPI definitions, Kiterunner can probably detect over 95% of endpoints.

Kiterunner is easy to use—specify the endpoint data file (a so-called Kite file) and run it against your target endpoint as shown here:

Figure 5.4 – Sample Kiterunner execution

The preceding Kiterunner scan was executed against a sample API application with both a `login` and `register` endpoint, and this was detected by Kiterunner as shown here:

```
POST    202 [     22,     4,    1] http://localhost:8090/api/register
0cc39f7908a098262e1a94331f0711f969c08024
POST    401 [     38,     4,    1] http://localhost:8090/api/login
0cc39f753802733afd682ea9673e6861e685415
```

Kiterunner can operate in *scan* mode using typical API endpoints derived from OpenAPI definitions, or it can operate in *brute* mode using more conventional word lists. Additionally, Kiterunner can be constrained to enumerate endpoints to a specified depth rather than to the full depth possible to improve scan times.

Fuzzing API endpoints

While Kiterunner functions by building a database of commonly encountered endpoints within OpenAPI definitions, other tools have a more rudimentary approach to *fuzzing* (defined as "*an automated software testing method that injects invalid, malformed, or unexpected inputs into a system to reveal software defects and vulnerabilities*"). Fuzzing tools will attempt to inject inputs into an API either within the path, parameters, or request bodies to create an unexpected result indicating a weakness. Fuzzing tools take a *word list* as input and then iterate through the values in this word list and record the responses.

Let us look at the *wfuzz* (`https://github.com/xmendez/wfuzz`) tool in operation against the same API used in the Kiterunner example:

```
docker run --rm -v $(pwd)/wordlist:/wordlist/ -it ghcr.io/
xmendez/wfuzz wfuzz -w wordlist/general/common.txt --hc 404
http://192.168.16.23:8090/FUZZ
```

For convenience, this example shows *wfuzz* being run in a Docker container using the following options:

- `run` tells Docker to run a container

- `--rm` tells Docker to remove the container upon termination

- `-v $(pwd)/wordlist:/wordlist/` mounts a local volume word list into the container

- `-it` specifies that an interactive terminal session is to be used

- `ghcr.io/xmendez/wfuzz` is the name of the container repository and image to be used

The rest of the command to be passed to the *wfuzz* executable provides the following options:

- `-w wordlist/general/common.txt` tells wfuzz to use the word list specified (this path is mounted in the container from the host OS).

- `--hc 404` tells wfuzz to suppress all responses with a 404 status code (these result from paths not found on the API).

- `http://192.168.16.23:8090/FUZZ` specifies the host, port, and path to be scanned. The FUZZ keyword is a positional parameter and will be replaced by each word from the word list as the scan progresses.

The execution of this command is shown in the following figure:

```
        # docker run --rm -v $(pwd)/wordlist:/wordlist/ -it ghcr.io/xmendez/wfuzz wfuzz -w wordlist/general/common.txt --hc 404 http://192.168.
16.23:8090/FUZZ
WARNING: The requested image's platform (linux/amd64) does not match the detected host platform (linux/arm64/v8) and no specific platform was requested
********************************************************
* Wfuzz 3.1.0 - The Web Fuzzer                         *
********************************************************

Target: http://192.168.16.23:8090/FUZZ
Total requests: 951

=====================================================================
ID           Response   Lines    Word      Chars      Payload
=====================================================================

000000025:   200        179 L    1486 W    12151 Ch   "about"
000000224:   301        10 L     16 W      173 Ch     "css"
000000413:   301        10 L     16 W      179 Ch     "images"
000000456:   301        10 L     16 W      171 Ch     "js"
000000489:   200        0 L      3 W       29 Ch      "login"
000000492:   302        0 L      4 W       28 Ch      "logout"
000000679:   200        0 L      3 W       29 Ch      "register"

Total time: 0
Processed Requests: 951
Filtered Requests: 944
Requests/sec.: 0

        #
```

Figure 5.5 – Sample wfuzz execution

In this example, wfuzz finds both the `login` and `register` endpoints found previously by Kiterunner.

Another popular fuzzer is Microsoft's restler-fuzzer (`https://github.com/microsoft/restler-fuzzer`), which is designed to take OAS definitions and scan APIs against the definitions to identify any vulnerabilities. Since there is a dependence on an OAS definition, restler-fuzzer is more likely to be useful in a development and test environment rather than for adversarial testing.

The key to successful fuzzing is to use a good set of word lists for your target—a fuzzer engine is only as good as its input. Fortunately, the InfoSec community has built a vast inventory of word lists for a number of different targets (such as host OS, platform, or service). The best resource for word lists is the master catalog produced by Daniel Miessler (`https://github.com/danielmiessler/SecLists`).

It is also worth noting that most fuzzing tools perform an equally good job at running password attacks against APIs and services. There are three primary forms of password attacks:

- **Brute-force attack**: The most basic form of attack iterates through a character set (alphanumeric and special characters) and submits as many passwords as possible before detection or limiting.

- **Dictionary attack**: A dictionary attack uses a word list of commonly used passwords and usernames and is likely to be more successful than a brute-force attack, but it requires a good word list.

- **Credential stuffing**: The most sophisticated attack uses a word list, but in this case, the list originates from credentials harvested from data breaches and compromises. Typically, these credentials can be obtained at a price on the *dark web* or via other hacker forums.

Fuzzing is a core skill in the armory of an aspiring attacker—make sure you understand the most appropriate attack type and use a good word list.

Attacking JWTs

The final item in your toolkit to attack APIs is the JSON Web Token Toolkit, aka jwt_tool (`https://github.com/ticarpi/jwt_tool`), which can be used for JWT testing, validation, tampering, and exploitation.

Running jwt_tool without any parameters will display the full details of the JWT provided, as shown here:

```
Version 2.2.6                                      @ticarpi

Original JWT:


[+] alg = "RS384"
[+] typ = "JWT"


[+] user_profile = JSON object:
    [+] id = 51
    [+] email = "user@acme.com"
    [+] password = "hellopixi"
    [+] name = "Pixi User"
    [+] pic = "https://s3.amazonaws.com/uifaces/faces/twitter/estebanuribe/128.jpg"
    [+] account_balance = 1000
    [+] is_admin = False
    [+] all_pictures = "[]"
[+] iat = 1667737223      ==> TIMESTAMP = 2022-11-06 12:20:23 (UTC)
[+] exp = 1667739023      ==> TIMESTAMP = 2022-11-06 12:50:23 (UTC)
[+] aud = "pixiUsers"
[+] iss = "https://42crunch.com"
[+] sub = "user@acme.com"


[*] iat was seen
[*] exp is later than iat by: 0 days, 0 hours, 30 mins
```

Figure 5.6 – Sample jwt_tool execution

jwt_tool can be used to verify a token against a public key in **Privacy Enhanced Mail** (**PEM**) or **JSON Web Key Set** (**JWKS**) format. From a hacker's perspective, the ability to attack JWTs and applications is the most useful—the tool allows a target URL to be specified, allowing the injection of JWTs into a target. Injection attacks can be automated using playbooks that attempt to exploit common JWT validation errors in a client application.

Finally, jwt_tool can be used to attack JWTs directly:

- By cracking the secret key for HMAC algorithms
- Using *key confusion* attacks against asymmetric ciphers with a known public key
- By forcing the use of the none algorithm to create unvalidated tokens
- By spoofing a JWKS

Now that we understand some of the low-level tools used by attackers to exploit vulnerabilities, let us now look at how to put these tools together to begin attacking APIs—first, let us look at API clients and proxying tools.

Mastering the tools of the trade

APIs are, by their nature, not exposed directly to the end user; rather, they are consumed via a mobile or web application, or perhaps via another API. To attack an API, we need to use a *client* and/or an *interception tool*, as discussed in the *Interacting with APIs* section.

The choice of tools is largely a personal one and my advice to you would be to choose one *client* (a **command-line interface** (**CLI**) such as curl or a GUI such as Postman) and one *interception tool* (such as Burp Suite) and become familiar with their usage across several scenarios.

CLI clients (HTTPie/cURL)

The simplest API client is a **CLI** client designed to be run interactively at a command prompt or terminal. They are particularly useful when testing connectivity to APIs or doing simple, quick debugging at the command line.

Most Unix-based OSs will come with either cURL (https://curl.se/) or wget (https://www.gnu.org/software/wget/) pre-installed or readily available from package managers.

I prefer the excellent *HTTPie* (https://httpie.io/cli) CLI client, designed specifically for use with APIs. It has many enhancements, such as the colorization of output, excellent JSON support, and great support for HTTPS, proxies, and authentication.

A sample *HTTPie* request is shown here; the flags specify that all headers should be shown and that TLS certificates should not be verified (useful for quick debugging):

```
,colind@mbm1pro: ~ # http -p=HBhbm --verify false http://localhost/uuid
GET /uuid HTTP/1.1
Accept: */*
Accept-Encoding: gzip, deflate
Connection: keep-alive
Host: localhost
User-Agent: HTTPie/3.2.1


HTTP/1.1 200 OK
Access-Control-Allow-Credentials: true
Access-Control-Allow-Origin: *
Connection: keep-alive
Content-Length: 53
Content-Type: application/json
Date: Sun, 06 Nov 2022 13:37:10 GMT
Server: gunicorn/19.9.0

{
}


Elapsed time: 0.034073125s

,colind@mbm1pro: ~ # 
```

Figure 5.7 – Sample HTTPie execution

Postman

An essential tool for API aficionados is the ubiquitous Postman (https://www.postman.com/), which is most easily thought of as a "browser, but for APIs."

While Postman is primarily intended as an aid for API developers (more on this in *Part 3, Defending APIs*), it has many features that make it an ideal tool for attackers wanting to explore and manipulate a target API.

Postman is available either as a web application or via a client for all popular OSs, and these environments can be synchronized via a platform account. Postman is free for personal use (and is suitable for any examples in this book), and has paid options for teams and enterprise users.

To understand how Postman works, let us look at some of the core elements in the user interface shown here:

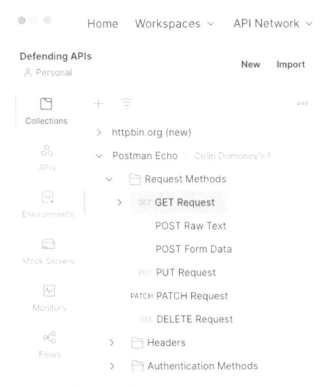

Figure 5.8 – Postman top-level navigation

Of interest to us are the top three tabs: **Collections**, **APIs**, and **Environments** (the other three options are used for testing APIs for functional and integration testing).

An *environment* in Postman is a collection of key-value pairs defining various configuration settings for a set of APIs being tested. Typically, this will include items such as the URL, port number, usernames, paths, tokens, and keys. By using an environment, it is easy to test the same API with a different set of conditions by simply switching the environment. Think of this as analogous to environment variables in a terminal CLI.

The **APIs** tab holds all APIs (defined via OpenAPI (versions 1.0, 2.0, 3.0, and 3.1), RAML (0.8 and 1.0), Protobuf (2 and 3), GraphQL, or WSDL (1.0 and 2.0)) within a Postman account. APIs can be created in one of three ways:

- Create a new API from scratch using the built-in API editor in Postman

- Import an external definition in one of the supported formats

- Use API requests captured in a *collection* as the starting point for the API definition

A typical API view is shown here:

Figure 5.9 – Postman API view

The **Collections** tab contains a group of API requests, including the request body, parameters, and authentication. A collection can be associated with an environment to inherit parameters, and collections can be published so that APIs can easily be shared via a central repository.

A typical collection view is shown here:

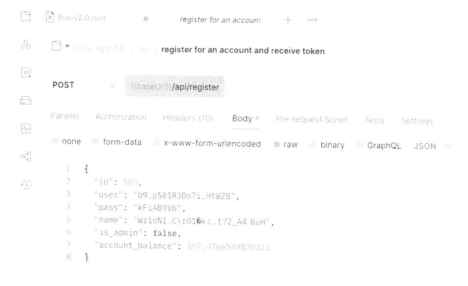

Figure 5.10 – Postman Collections view

Postman also can generate client code for many languages, including the popular CLI tool cURL for API debugging, as shown here:

Figure 5.11 – Postman code snippet generation

We have only scratched the surface of some of the features in Postman; other powerful features include a proxy server, command-line test runners, mock servers, and full proxy support. This should be the go-to tool for any serious API developer or hacker.

Browser tools

One of the most useful ways to discover APIs and explore their functionality is to use a web page of the site of interest. Increasingly, websites are built using **single-page application** (**SPA**) patterns or leverage asynchronous calls to backend APIs to retrieve data on demand. All modern browsers support powerful built-in developer tools targeted at frontend developers to aid in web development. These tools can also be useful to API attackers using powerful network inspection tools to monitor backend API traffic from within the browser without using external intercepting proxies (and the corresponding hassle of installing client certificates and switching the browser proxy settings).

As an example, let us assume we are trying to understand how the LinkedIn API works. By navigating to our home profile page and observing the backend connections, we can see an API call as shown here:

Figure 5.12 – Browser tools network panel

By examining the API request and response headers and body, an attacker can glean a lot of useful information about the API. For instance, in this example, the response data body shows some interesting metrics, including connection count. Once an attacker understands the API endpoints and request parameters, they can switch to more sophisticated methods to attack and exploit the API.

The following screenshot shows the wealth of information revealed by the `relationship/connectionsSummary` endpoint:

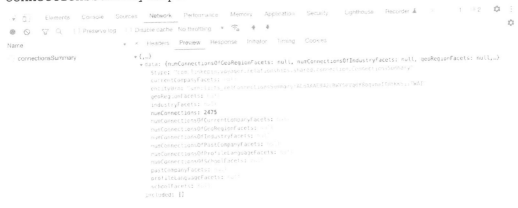

Figure 5.13 – View API data in network tools

As shown, I have a total of 2,475 connections on LinkedIn. Although this API is not officially published or supported by LinkedIn, it is readily accessible publicly and could easily be abused by an adversary.

Burp Suite

We will now look at another essential tool (with Postman being the other essential tool, discussed earlier) for any aspiring API attacker—the legendary **Burp Suite** (`https://portswigger.net/burp/communitydownload`) web security testing tool. Since its origin in 2003, Burp Suite has been the principal tool used by serious web application penetration testers. Burp Suite is available for download and has several different licensing models: Enterprise (with support for the automated scanning of web applications), the affordable Professional version, and the free Community edition, which will be used in all examples in this book.

We will take a brief look at some of the key features of Burp Suite; more detailed examples will be shown in *Chapter 7*. The core component of Burp Suite is the proxy engine, which allows Burp Suite to intercept HTTP traffic, process it via several internal engines, relay it to the server, and then process the response before relaying it to the client. This is best understood by looking at *Figure 5.2*.

The proxy allows advanced configuration of its behavior, including options such as certificate management, filters, match and replacement, and payload modification. The most useful feature is the intercept capability, which allows a user to intercept the request and inspect it in the user interface and either drop the request, forward it, or modify it and then forward it. An example of the intercept capability is shown here, which shows a request captured in Burp Suite before being forwarded to the client (**Forward**) or ignored (**Drop**):

Figure 5.14 – Burp Suite intercepting proxy

This intercept feature is particularly powerful for attackers since it allows them to inspect the normal traffic to and from a target. Then, once they have identified a weakness, they can modify the request and observe the resultant behavior.

Once a target has been probed via the proxy, Burp Suite will build up a site map of the target, indicating URLs, methods, status codes, and request and response bodies. The site map is a convenient navigation point for further exploration using the other engines. The following figure shows a sitemap for the Pixi API application:

Figure 5.15 – Burp Suite site map view

The processing engines included in the Community edition include the following:

- **Repeater**: As the name suggests, Repeater allows the user to manipulate and re-issue multiple requests to the target to attempt to identify unusual behavior.

- **Intruder**: This is probably the most powerful tool and one we will examine in detail.

- **Target**: This includes the site map shown previously and allows the user to bookmark and highlight requests of interest.

- **Logger**: The engine can store detailed logs of all traffic being processed by Burp Suite, either for further analysis in Burp Suite or via export to third-party tools.

- **Sequencer**: This engine is useful for analyzing a target's response to randomized data such as password reset tokens and anti-CSRF tokens.

- **Comparer**: This tool provides the visual difference between two data samples, and it is typically used to compare and contrast two different transactions and to easily show the differences.

- **Decoder**: Finally, this tool converts input data into a human-readable format and supports common formats such as URLs, HTML, Base64, ASCII hex, Hex, Octal, binary, and Gzip.

To understand the power of Burp Suite, let us look at the **Intruder** engine being used to crack an API authentication password. Firstly, **Intruder** is configured with a payload to attack the target. In the following example, I am using a very small manually entered payload of five entries; however, any publicly available word list can be used in the payload.

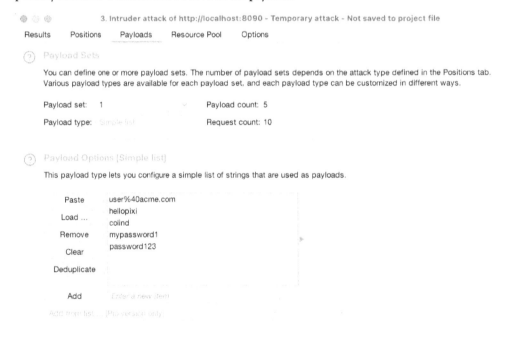

Figure 5.16 – Burp Suite Intruder payload configuration

The next step is to sweep all possible combinations of the payload against the target and observe the responses. In this example, we can see that a combination of hellopixi and user@acme.com resulted in a 200 OK status code and the return of a JWT. This indicates a successful login, while all the other responses resulted in a 401 Unauthorized status code:

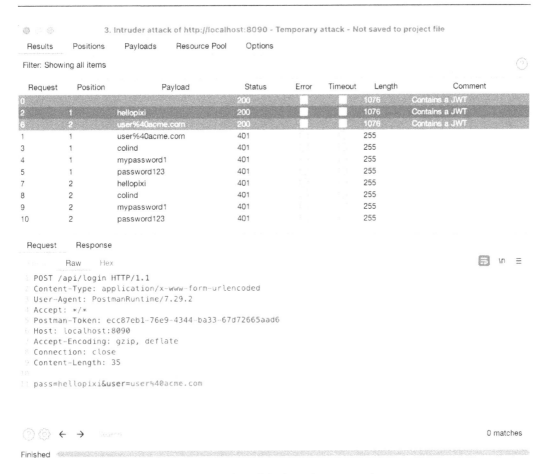

Figure 5.17 – Burp Suite Intruder scan results

In that event, the already impressive built-in functionality is missing something; there is almost certainly a plugin available in the Burp Suite marketplace (the BappStore) and community to meet that need. Of particular interest to API attackers are the following:

- **IP Rotate**: This plugin uses multiple API gateways across regions to randomize the source IP address for requests. This reduces the likelihood of triggering either **Intrusion Detection Systems (IDS)** or rate-limiting / throttling of APIs.

- **JWTs**: The ubiquity of JWTs in APIs means that attackers frequently need to inspect and verify the JWTs. This plugin detects the presence of a JWT within a request or response and provides a graphical view of the content—which is invaluable for evaluating claims in tokens.

The JSON Web Tokens extension is shown in detail here:

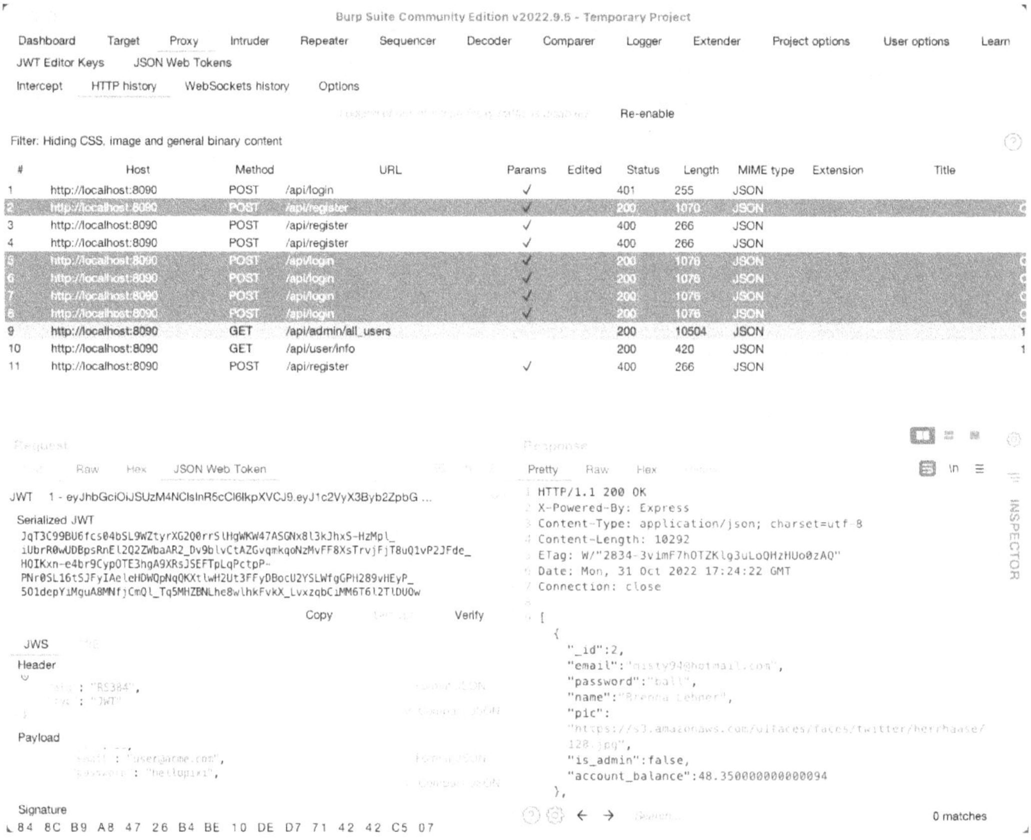

Figure 5.18 – JSON Web Tokens extension

An excellent guide to some of the best Burp Suite extensions is available in the excellent reference to awesome Burp Suite extensions in the *Further reading* section.

Finally, the team at PortSwigger provides some of the most comprehensive training on Burp Suite, and also the full spectrum of web and API security topics, including many interactive laboratories and hands-on exercises:

```
https://portswigger.net/web-security/dashboard
```

Reverse proxies

The final tool that we will cover is a very useful and ubiquitous reverse proxy application, mitmproxy (`https://github.com/mitmproxy/mitmproxy`), which provides similar reverse proxy features to Burp Suite. The real benefit of mitmproxy, though, is the fact that it is lightweight, can be run at the command line, and has a package or port for nearly every OS. As their website describes it, it is the Swiss Army knife for debugging, testing, and pentesting.

While mainly used from the command line, it also offers a user interface similar to the Chrome developer tools, and it provides a Python API so curious developers can write add-ons and extensions.

For API developers, mitmproxy provides another particularly powerful feature—the ability to reverse engineer an OpenAPI definition for an API based on observer web traffic. Using mitmproxy2swagger (`https://github.com/alufers/mitmproxy2swagger`), a user can export a traffic capture from mitmproxy and convert it into an OAS definition. This is very useful if you are working with an API that does not have an OAS definition.

The following screenshot shows four different API transactions captured in mitmproxy:

Figure 5.19 – mitmproxy API traffic capture

The captured traffic is exported and, using mitmproxy2swagger, a YAL-format OpenAPI definition is produced for the API under observation, as shown here:

```
pixi.yml U ✕

pixi.yml > {} paths
      Validation schema for OpenAPI Specification 3.0.X  (v3.json)
 1    openapi: 3.0.0
 2    info:
 3      title: flows Mitmproxy2Swagger
 4      version: 1.0.0
 5    servers:
 6    - url: http://localhost:8090/api
 7      description: The default server
 8    paths:
 9      /login:
10        post:
            Scan | Try it
11          summary: POST login
12          responses:
13            '200':
14              description: OK
15              content:
16                application/json:
17                  schema:
18                    type: object
19                    properties:
20                      message:
21                        type: string
22                      token:
23                        type: string
24      /user/info:
25        get:
            Scan | Try it
26          summary: GET info
27          responses:
28            '200':
29              description: OK
30              content:
31                application/json:
32                  schema:
33                    type: array
34  >                 items: ⋯
54      /admin/all_users:
55        get:
```

Figure 5.20 – mitmproxy2swagger generation OAS definition

Now that we understand the tools we will use to attack APIs, let us go onto the exciting part—putting those skills to use.

Learning the key skills of API attacking

The best way to learn how to hack APIs is by getting hands-on experience with hacking tools and a vulnerable API. We will start by building our own API hacking laboratory using many of the popular tools we have explored, and then use this laboratory to attack some deliberately vulnerable API-based applications. This hands-on approach will equip you with the skills necessary to attack real-world APIs in the upcoming chapters.

Building a laboratory

The choice of a personal laboratory is somewhat dependent on individual preferences for OSs and favored tools. For most users, I would recommend three core components: a versatile modern editor (such as Visual Studio Code), a container runtime environment (such as Docker), and the ubiquitous Kali Linux hacking OS.

Visual Studio Code

The choice of a code editor and/or **integrated development environment** (**IDE**) is a deeply personal one (as anyone who remembers the *vi versus emacs* debates of yore will know). My preference is for the excellent **Visual Studio Code** (`https://code.visualstudio.com/`) from Microsoft. This fully open source editor is built to be fast, responsive, and highly extensible through an open marketplace. It works perfectly out of the box, but can easily be customized and tweaked to your heart's content. In addition, for API developers and hackers, there are some excellent plugins I would recommend:

- **Thunder Client** (`https://marketplace.visualstudio.com/items?itemName=rangav.vscode-thunder-client`): This is a lightweight API client with a user interface similar to Postman. I have found this to be useful for running quick calls to APIs to observe their response.

- **OpenAPI (Swagger) Editor** (`https://marketplace.visualstudio.com/items?itemName=42Crunch.vscode-openapi`): This extension from 42Crunch provides syntactical highlighting of OpenAPI definitions, recommendations for best practices, and code completion and generation features. Additionally, the plugin integrates with the 42Crunch API security platform to provide API scan and protection features.

Docker

Docker (`https://www.docker.com/get-started/`) has become *the* essential tool for containerized software development. A container can be thought of as a self-contained application and runtime environment partitioned from the rest of the host OS—conceptually, it is as if your application is running on its own with its own execution environment, filesystem, memory, libraries, and so on.

The major benefit from a development viewpoint is encapsulation—no longer does a developer have to ship an application with an installer program or struggle with package or dependency conflicts in the target environment. Instead, they ship their application with all its dependencies and configuration in a container, and on the target system, Docker can execute the container. And if you discover you need more instances of your application—just run another few instances of the container; this is the concept of scaling. If this all gets too complicated to manage, you can use a container orchestrator (such as Kubernetes) to manage your containerized application(s).

Many of the tools and examples we will use in this book are packaged as Docker containers for convenience and shareability. If you are going to be working on the examples, you should make sure you have Docker installed on your development machine, and that you have a basic understanding of how to use containers.

Kali Linux

As much as Burp Suite is the de facto hacker's tool of choice, **Kali Linux** (`https://www.kali.org/get-kali/`) is their OS.

Kali Linux is a hardened Debian-based, open source OS targeted at penetration testers and security auditors. It comes pre-installed with over 600 penetration testing tools (including nearly all the ones covered in this chapter), supports a wide range of wireless devices, and is developed in a secure environment by a small team of dedicated professionals.

Kali Linux is also available in various form factors, ranging from bare-metal installations to virtual machine images, to embedded platforms such as Raspberry Pi. For the purposes of this book, a virtual machine instance of Kali Linux is perfectly adequate and we will not be using any bespoke wireless network devices.

Once you have Kali Linux installed and running, ensure you perform a full system update to retrieve the latest versions of packages and applications.

Hacking vulnerable APIs

Now that you have all the hacking tools at your disposal, you need a (vulnerable) target to attack. Of course, you could start attacking live targets such as public APIs; however, this may land you in hot water with the owners or in contravention of your local internet regulations. Also, most public APIs will be relatively free of vulnerabilities, making your learning experience challenging.

By far the best option is to use one or more of the many excellent deliberately vulnerable API-based applications available online. A few of my top choices are included here:

Project Name	URL	Comment
crAPI	`https://github.com/OWASP/crAPI`	**Completely Ridiculous API (crAPI)**
Pixi	`https://github.com/DevSlop/Pixi`	Vulnerable REST API based on the MEAN stack
VAmPI	`https://github.com/erev0s/VAmPI`	OWASP API Top 10 vulnerabilities
vAPI	`https://github.com/roottusk/vapi`	OWASP API Top 10-based exercises
BankGround	`https://gitlab.com/karelhusa/bankground`	Open source project to learn REST and GraphQL security
HTTPBin	`https://httpbin.org/`	Simple HTTP request and response service

Table 5.1 – Recommended vulnerable API applications

Many of the examples in this book are based on the Pixi application; this is a great starting point.

Training courses

If you prefer a more academic approach to learning about API security, then my first recommendation is the excellent content from Philippe De Ryck on a range of API security, secure development, and authorization and authentication topics. Full course details are available on his website: `https://courses.pragmaticwebsecurity.com/courses/api-security-best-practices`.

Other online content is also available from Pluralsight (`https://www.pluralsight.com/courses/owasp-top-ten-api-security-playbook`) and AppSecEngineer (`https://www.appsecengineer.com/courses-collection/api-security-attack-and-defense`).

Finally, if you want to take your skills to another level, then the *API Security University* course is the best choice; more details can be found here: `https://university.apisec.ai/`

The course is authored by Corey Ball (author of *Hacking APIs*) and features laboratory examples, vulnerable code, and a vibrant Discord community.

Summary

We have covered a lot of ground in this chapter, and you are well placed to begin the next step of our journey—discovering and attacking APIs. The first consideration when attacking an API is how to interact with it (usually via a reverse proxy), followed by gathering metadata about the API, including keys, tokens, and endpoints.

We learned how API hackers are spoilt for choice when it comes to tools to use against APIs. By far the most important of these are the Postman API browser and the Burp Suite security testing tool. Finally, we covered several excellent educational resources available to API hackers.

Let's dive into the next exciting chapter in our journey—looking at how to discover APIs.

Further reading

- The best guide to API security resources: `https://github.com/arainho/awesome-api-security`

- Awesome Burp Suite extensions are available here: `https://github.com/snoopysecurity/awesome-burp-extensions`

- Assorted API security training courses and playlists:

  - `https://www.youtube.com/playlist?list=PLbyncTkpno5HqX1h2MnV6Qt4wvTb8Mpol`

  - `https://www.wallarm.com/what/how-to-hack-api-in-60-minutes-with-open-source`

  - `https://hackanythingfor.blogspot.com/2020/07/api-testing-checklist.html`

  - `https://www.wallarm.com/what/how-to-hack-api-in-60-minutes-with-open-source`

- Articles from Approov.io on how to perform MitM attacks on mobile applications:

  - `https://approov.io/blog/how-to-mitm-attack-the-api-of-an-android-app`

  - `https://approov.io/blog/securing-https-with-certificate-pinning-on-android`

6
Discovering APIs

In the previous chapter, we explored the foundations of attacking APIs, focusing on many of the tools that attackers use. In this chapter, we'll use these skills to learn how to discover APIs in the real world. We will learn how to discover APIs using passive methods (where we do not interact with the API directly) and active methods (where we interact with the API directly). We will also learn how to find details of how the API is implemented and how to use this knowledge to attack an API.

For an API defender, it is important to understand the techniques used by your adversaries in discovering your APIs so that you can implement defensive measures to prevent easy discovery and further analysis. In particular, defenders should pay attention to attackers' techniques in identifying implementation details and use this knowledge to harden their implementations.

For an API attacker, a thorough reconnaissance process provides useful information about target APIs and their implementation. The more thorough the discovery phase, the greater the likelihood of a successful exploit phase. The API documentation provides a wealth of information regarding the API implementation and how this might be attacked, particularly concerning authentication and authorization.

In a nutshell, this chapter is going to cover the following main topics:

- Passive discovery
- Active discovery
- Implementation analysis
- Evading common defenses

Technical requirements

For this chapter, you will need a development machine capable of the following:

- Running Docker locally

- Running VS Code with various marketplace extensions

- Access to the internet and a GitHub account to access the examples

This chapter contains many code samples in various languages; these can either be run locally, which will require the installation of compilers, SDKs, and frameworks, or from within a Docker build container.

The example code and various breaking changes to the instructions can be found in the `Chapter 6` folder on the book's GitHub repository here: `https://github.com/PacktPublishing/Defending-APIs/tree/main/Chapter6`

Passive discovery

In the first section of this chapter, we will investigate various methods of the passive discovery of APIs. Not surprisingly, the techniques involve utilizing different search engines or online repositories to mine useful API metadata.

The finer details of a passive discovery phase will be determined by the target in mind. Sometimes the intent will be to search *far and wide* (for example, try and identify all APIs that an organization owns) and gain as much information as possible about the number of exposed targets available on the public internet. For example, let's imagine you are attempting to exploit the API of a new router with a vulnerability—in this case, you will probably be attempting to find online instances with public IP addresses.

In other scenarios, you may know about a particular exploit and want to gain a deeper knowledge of using the exploit in practice. You might use a more *narrow and deep* strategy (for example, identify only the APIs using OAuth2 for authentication) to gain information about specific endpoints and behaviors.

Google

It will probably come as no surprise that Google is the first recommended resource for API discovery. This behemoth of a search engine indexes much of the world's connected computing resources, and its APIs provide a wealth of metadata that can be indexed. Other search engines are available, but this section will focus on Google and its derivatives.

Simply typing a query term into Google can produce useful results. For example, searching *API for Covid-19 tracking* returns several pages of sites or services that host Covid-19 tracking APIs. Probably the most useful refinement that can be made to a search is using the various *query operators* available via the advanced search mode on Google. By appending one or more operators, the result set can be refined to allow very specific searches. Some of the most used search operators are shown here:

Query operator	Search result modification
Intitle	Searches for a term in the site title
Allintitle	Searches for all terms in the site title
Inurl	Searches for a term in the site URL
Ext	Searches for files with the specified extension
Intext	Searches for a term in the search results
Allintext	Searches for all terms in the search results
Site	Restricts the scope of the search to the site specified

Table 6.1 – Common Google query operators

The query operators are best understood using an example. Imagine, as an aspirant API attacker, you want to learn more about the Twitter API, particularly the *OAuth2* flow. You know that the API uses *OAuth2*, but you do not have further details. By using a simple query, such as `site:api.twitter.com inurl:/callback`, it is easy to discover two endpoints, as shown:

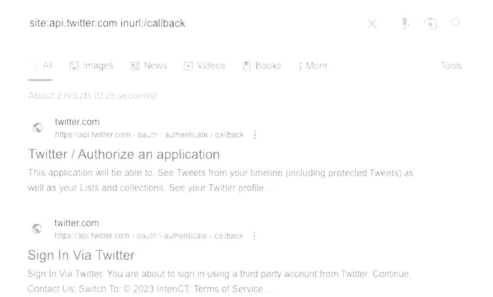

Figure 6.1 – Using Google query parameters to discover login URLs

The *Further reading* section provides additional resources for using these Google query operators, frequently termed **Google dorks** in the industry.

Offensive security Google database

In fact, the use of Google dorks led the team at **Offensive Security** (the maintainers of the Kali Linux **operating system** (**OS**) covered in *Chapter 5, Foundations of Attacking APIs*) creating a dedicated website for popular hacking resources. The **Google Hacking Database** (**GHDB**) includes a searchable index of community Google dorks (`https://www.exploit-db.com/google-hacking-database`). Filtering by the search term `api` reveals some interesting results, as shown in the following table:

Query	Scope of search
`Intitle:"index of" api*.txt`	List possible exposed API keys in file indexes
`inurl:pastebin "API_KEY"`	Search the Pastebin repository for `"API_KEY"`
`intext:api_key filetype:log`	Search log files for `"api_key"`
`allintext:"API_SECRET*"` `ext:env | ext:yml`	Search env and yml files for `"API_SECRET*"`

Table 6.2 – Selected GHDB API queries

Google dorks are also useful for finding exposed devices online, such as Wi-Fi routers and webcams. Take some time to explore the GHDB and understand how the queries work, and use them to search for some exposed, valuable API information.

Other API-specific searchable databases

Although Google and the GHDB are the biggest online information resources for passive API reconnaissance, many specialist databases or indexes of APIs are worth investigating during the research phase.

The original (and for many years the only) API database online was the veritable **ProgrammableWeb** (`https://www.programmableweb.com/`) founded as far back as 2005 and only shut down in February 2023. For the last decade, MuleSoft had owned ProgrammableWeb and promoted it as the "*go-to destination for APIs and Integration.*" At the time of its shutdown in 2023, the site listed over 23,000 APIs, including their endpoint details, version information, API status, documentation, and even SDKs.

This section will examine a few of the most popular alternatives to ProgrammableWeb.

APIs.guru

APIs.guru (`https://apis.guru/`) is an online directory of 2,518 APIs (at the time of writing) with a specific focus on ensuring the quality of content. For example, they filter out private APIs that are found to be unreliable. They require that submitted APIs are in one of the common API specification formats (ideally OpenAPI 3.x or Swagger) and encourage submitters to remediate issues within their specifications. Finally, APIs.guru has its own API, which allows users to query public APIs easily.

Public-APIs GitHub repository

One of the largest repositories of APIs is the **public-apis** GitHub repository (`https://github.com/public-apis/public-apis`), which contains several thousand public APIs sorted by category with direct links to the provider sites. Although extensive, there do appear to be some data quality issues with some of the entries. At the time of writing, it seems the project is stagnant; however, this may change in the future.

Unfortunately, this repository does not have an API or a queryable interface, making searching cumbersome.

APIsList

APIsList (`https://apislist.com/`) is a relatively modern API database featuring a searchable interface and well-organized categories. Users can filter by APIs that are accessible without an API key or with **Cross-Origin Resource Sharing** (**CORS**) support, or are open source. When browsing the API details page, there are links to the API documentation page, a live demonstrator of the API (if available), and suggestions for similar APIs.

Rapid API

The emergent replacement for ProgrammableWeb appears to be **Rapid API** (`https://rapidapi.com`), which has a very impressive catalog of APIs sorted by category and is searchable. Estimates vary on the usage figures of Rapid API; however, recent reports suggest that over a million developers have tested or used APIs hosted by them. The site also offers statistics regarding APIs, including their popularity, latency, and service level.

The site has a distinctly commercial feel due to the many commercial APIs hosted on it, and it also offers a subscription plan (starting with a free tier) that allows end users to test a commercial API. Think of Rapid API as a broker service for APIs. This can be very useful in testing out an API that may be otherwise inaccessible, such as the Twitter v2 API, shown in the following screenshot:

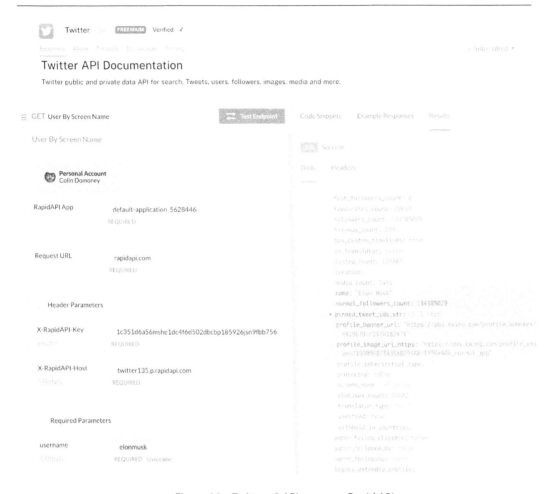

Figure 6.2 – Twitter v2 API query on Rapid API

In this example, using my free account on Rapid API, I can query the Twitter v2 API (inaccessible to end users at the time of writing) via the Rapid API portal.

Postman Network

The other challenger for the ProgrammableWeb crown is **Postman Network** (https://www.postman.com/explore), provided by Postman (its API browser tool was covered in *Chapter 5*, *Foundations of Attacking APIs*). Given the enormous popularity of Postman as an API development tool, it is no surprise that many developers opt to host their APIs on Postman. API-first organizations wishing to drive the adoption of their APIs are also driven to publish their APIs on Postman Network. This has resulted in dramatic growth; currently, the network has 2.5 million teams, 140k workspaces, 21.8k APIs, and 307k collections.

The key attraction for API developers is that Postman Network offers many developer-oriented features, such as version control, branching/forking, mock servers, and documentation. Additionally, API owners can opt to make APIs public or privately accessible to only their team(s).

Shodan

The final search tool we will focus on is the somewhat infamous **Shodan** (`https://www.shodan.io`), the de facto tool for searching devices connected to the internet. The easiest way to understand what Shodan does is to consider it a Google for internet-connected devices rather than for web content. While Google will index pages and their responses, Shodan will index the TCP/IP protocol responses for all IP addresses and ports it can discover. Conceptually, it builds a map of the internet, cataloging all responses as it crawls the internet.

Shodan analyzes the responses by analyzing the *banner* (the response headers) and indexing the response. Like Google query operators, Shodan has a query language that allows queries to be run against many metrics. Security researchers primarily use Shodan to find vulnerable devices on the internet. Searching by vulnerability disclosure using the **vuln** operator is possible in the professional subscription.

In performing reconnaissance on APIs, Shodan can be extremely useful in narrowing down a target for attack. For example, by restricting `content-type` to either `application/json` or `application/xml`, the search can be narrowed to endpoints that are most likely to return API data.

Let us take an example where we run a query, `"content-type: application/json" product:"Kong Gateway" version:"2.1.4" "200 OK"`, which searches for all JSON response data with a `200 OK` status hosted on Kong Gateway version 2.1.4. Taking a closer look at the first result (of the seven returned) shows us a wealth of information, as shown in the following screenshot:

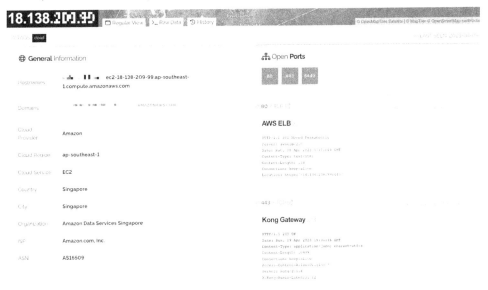

Figure 6.3 – Shodan detailed results view

At this point, we almost certainly have an API endpoint and can use other tools (such as Kiterunner) to discover and evaluate the endpoints of this API.

Curious readers may be interested in running a Shodan search on their own IP address—I certainly found some illuminating information while writing this book!

Code analysis techniques

The previous techniques have focused on the host or network layer of discovery. It is now time to turn our focus to passive discovery using source code analysis techniques.

GitHub

As the world's largest source code repository, **GitHub** is the obvious place to go to try and identify and discover APIs by examining source code. Any public repository on GitHub is indexable by Google and a number of third-party search tools.

First, let us use our Google dorks techniques from earlier in this chapter to find code that appears to be hosting APIs in .NET Core. By convention, API endpoints are often implemented in files with the name `Controller` in their path and will have the `cs` extension. So, a Google search with the term `site:github.com ext:cs Controller` should produce results, and indeed we find over 7,300 files worth further exploration.

While the preceding searches are useful, they lack insight into the code itself. Fortunately, there are excellent third-party tools that add search capabilities to the source code—let us take a look at the popular **grep.app** (`https://grep.app/`), which claims to index all GitHub public repositories. Using our previous use case, let us see how `grep.app` can search for code that appears to be hosting an API endpoint in .NET Core. In this case, we can make the search much more specific since we know that any API endpoint in .NET Core taking a `POST` request will have an `[HttpPost]` decorator applied to the method. Using `grep.app` to perform a search, we get the following results:

Figure 6.4 – Using grep.app to search GitHub

This search shows how easily we can locate .NET Code source files with the `[HttpPost]` decorator. The search can also be restricted to a specific language, their repositories, and the file path.

Examining JavaScript files

In my experience, the richest harvest of API client and server code is to be found with JavaScript-based languages since these tend to be **isomorphic** (they can run on the client, in the browser, or on the server). The vast majority of APIs will feature a client (mobile application or browser-based) written in JavaScript, or provide a client SDK in JavaScript. Knowing how to hunt for API code in JavaScript is a key skill for a budding API hacker.

The first code pattern to search for is the common methods and functions for making API calls. Typical favorites here include the `XMLHttpRequest` method, the jQuery `$.ajax()` method, or the popular `axios` library. Using `grep.app` to search for `axios.get` reveals nearly 20,000 hits. As discussed in the section on Google dorks, search for anything that might be an API key or `token.` `grep.app` finds a total of 154,000 instances of `API_KEY`.

JavaScript code is usually built or packaged with a package manager such as **npm**, and this leaves telltale signs of the nature of the code; for example, is it including an API client such as `axios` in its `package.json` file? Configuration and environment files may hold all sorts of useful API-related information, such as paths and URLs.

One of the best places to look for JavaScript API code is within the *developer tools* of your browser, and that is where we will go next as we move on to methods for the active discovery of APIs.

Active discovery

We will now look at various techniques for the **active discovery** of APIs in the real world. By active, we mean that we will interact with the API and/or its network by monitoring the traffic or directly accessing the API or its host.

> **Note – ensure that you have permission to access the computing resources**
>
> The use of active discovery using the tools and techniques described in this section may be against the terms and conditions of use of various services, ranging from your ISP to the relevant cloud hosting service. If you are unsure whether you are authorized to perform such scanning, you should err on the side of caution and seek explicit permission. In many cases, such scanning may also be against the applicable laws within your country, and any violations may have serious consequences. Fortunately, many of the scenarios described can easily be recreated in a laboratory environment that is totally under your control.

Network discovery and scan

Typically, the first challenge to be overcome when trying to perform the active discovery of APIs is to know the range of IP addresses used by their host infrastructure. The IP address range depends on the infrastructure in question. In a closed environment, the IP address range may be an **RFC 1918** (https://www.rfc-editor.org/rfc/rfc1918) private range and well known; in other cases, an ISP or a cloud provider may own the range.

For an ISP, its IP address range can be obtained by finding its **Autonomous System** (**AS**) number and using this with a *whois* query (either locally on the command line or via an online tool) to get the IP address range.

For a cloud provider, there are various techniques to use to determine the IP address range of its compute resources. For example, in the case of AWS, use passive discovery tools to determine AWS-hosted domains and its ranges. Alternatively, try and identify AWS API Gateway endpoints or other AWS-hosted APIs, such as Elastic Beanstalk, EC2 instances, or ECS containers.

Now that you have identified a candidate IP address range, it is time to scan for candidate hosts serving an API. Let us examine two popular tools for this purpose.

nmap

The industry stalwart for network scanning is the legendary **nmap**, which is celebrating over a quarter of a century in use. nmap sends TCP/IP packets and transactions and then analyzes the responses to determine the host information. Typically, nmap works in one of four major modes, namely the following:

- **Host discovery**: It checks which hosts exist in a given IP address range
- **Open port scan**: It determines which ports are open for a given host
- **Service discovery**: It determines which services are running for a given host, by analyzing their responses to probing
- **Test for vulnerabilities**: It uses scripting to search for common vulnerabilities

For the purposes of API discovery, it will be the first two capabilities that are most useful: identify which hosts are running a service on ports 80, 443, and other common HTTP ports.

First, let us look at how to scan a /24 CIDR address range for hosts that are alive using the following command:

```
colind@mbm: ~ # sudo nmap -sn 192.168.9.0/24
Starting Nmap 7.93 ( https://nmap.org ) at 2023-04-13 17:54 BST
Nmap scan report for 192.168.9.1
Host is up (0.012s latency).
MAC Address: 00:1D:AA:A6:DC:F8 (DrayTek)
Nmap scan report for 192.168.9.13
Host is up.
Nmap done: 256 IP addresses (2 hosts up) scanned in 1.98 seconds
```

This response identifies two hosts: the first is a router on the 192.168.9.1 address, and then my laptop is on the 192.168.9.13 address.

Next, let us use nmap to identify the open ports and running services on my laptop using the following command:

```
sudo nmap -p- 192.168.9.13
Starting Nmap 7.93 ( https://nmap.org ) at 2023-04-13 18:18 BST
Nmap scan report for 192.168.9.13
Host is up (0.00047s latency).
Not shown: 65527 closed tcp ports (reset)

PORT      STATE SERVICE
22/tcp    open  ssh
445/tcp   open  microsoft-ds
5000/tcp  open  upnp
5900/tcp  open  vnc
```

```
7000/tcp   open   afs3-fileserver
13231/tcp  open   unknown
17500/tcp  open   db-lsp
54399/tcp  open   unknown
Nmap done: 1 IP address (1 host up) scanned in 6.81 seconds
```

This response identifies several services on my laptop, including SSH, VNC, and the AFS3 file server.

There is much more to nmap than shown here; it should be the tool of choice for determining what is running on a network segment.

Masscan

The other scanner worth a call-out is the **Masscan** project (https://github.com/robertdavidgraham/masscan), billed as an *internet-scale port scanner*. The scanner's *raison d'etre* is to scan wide and very fast, with the project owner Robert David Graham claiming that it can scan the entire internet in under five minutes. It supports many of the same modes of operation as nmap, and utilizes its own TCP/IP stack to provide advanced capabilities and performance.

I used Masscan to scan my home laboratory and found it extremely performant and accurate in finding all known hosts, as shown in the following example:

```
# sudo ./bin/masscan -p80,8000-8100 192.168.16.0/24
Starting masscan 1.3.2 (http://bit.ly/14GZzcT) at 2023-04-13 19:19:37
GMT
Initiating SYN Stealth Scan
Scanning 256 hosts [102 ports/host]
Discovered open port 80/tcp on 192.168.16.6
Discovered open port 8080/tcp on 192.168.16.6
Discovered open port 80/tcp on 192.168.16.1
Discovered open port 8069/tcp on 192.168.16.16
Discovered open port 80/tcp on 192.168.16.16
Discovered open port 80/tcp on 192.168.16.5
Discovered open port 8069/tcp on 192.168.16.1
Discovered open port 8069/tcp on 192.168.16.17
Discovered open port 8022/tcp on 192.168.16.1
Discovered open port 8081/tcp on 192.168.16.6
Discovered open port 80/tcp on 192.168.16.17
```

In the words of the project owner: *"Masscan is tuned for wide range scanning of a lot of machines, whereas nmap is designed for intensive scanning of a single machine or a small range."*

OWASP ZAP

Now, having identified one or more hosts with open ports, it's time to take a deeper dive into what is running on the host. There are several tools to accomplish this, but let us take a look at the **OWASP ZAP** project (`https://www.zaproxy.org/download/`) dynamic scanner. One of the most useful features of ZAP is the ability to run a so-called *spider scan* (so-called because the scan *spiders* out to the endpoints on a host) on a host/port combination to discover details of the server and application.

To illustrate this powerful capability, the following example shows OWASP ZAP executing a spider scan against a locally installed WordPress server:

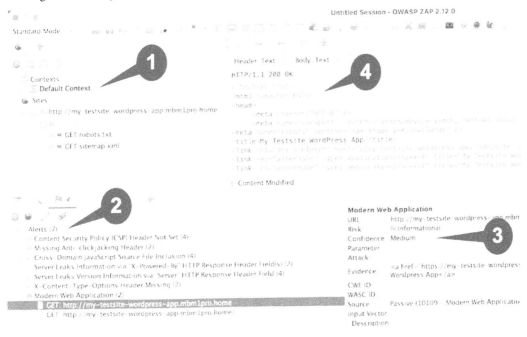

Figure 6.5 – OWASP ZAP running a spider scan

The four main panels provide the following information:

1. The site layout pane shows the root of the site and the hierarchy.

2. The results panel shows detailed information on the findings. In this view, it shows alert-level notifications.

3. The details panel shows deeper-level information about the selected alert.

4. The raw source view displays the response received, in this case, an HTML payload.

ZAP can identify a wealth of information in a very brief scan, including the directory structure, implementation details, and possible vulnerabilities.

Burp Suite

We covered **Burp Suite** in detail in *Chapter 5, Foundations of Attacking APIs*, and covered the ways in which it can be used to augment a manual test; however, it is worth mentioning that the paid versions of Burp Suite (*Professional* and *Enterprise*) also offer the capability to perform spidering of target sites to produce a similar sitemap for further analysis.

Reverse-engineering mobile apps

In *Chapter 5, Foundations of Attacking APIs*, we showed how a reverse proxy tool could intercept communication between a web application and its server. Similarly, a reverse proxy can be used to intercept traffic from a mobile application. The steps required are generally similar, namely as follows:

1.   Install the reverse proxy root certificate on your mobile device.

2.   Set the reverse proxy to accept inbound proxy connections on a known port, for example, 8888.

3.   Set the mobile device to use a proxy and set this to your host IP address and the port selected.

4.   Use the mobile application as normal in regular usage and record the proxied traffic.

Many modern mobile applications employ **certificate pinning** to ensure they are impervious to such simple **machine-in-the-middle** attacks.

Postman

Finally, the **Postman API platform** comes replete with a reverse proxy capability capable of intercepting API traffic and displaying it within a *Postman session* (with the option to save captured sessions to *Postman collections*). Additionally, Postman provides the Interceptor plugin, which runs within Chrome and can capture traffic from Chrome sessions within Postman for further analysis, replay, or hacking.

We have now learned how to use both passive discovery to identify APIs from public information sources and active discovery techniques to examine their behavior.

Implementation analysis

Finally, we will conclude this chapter with some tips on how to glean additional information about the implementation of the API's server, including the host OS and the libraries and frameworks used, including version numbers. Such information can be immensely useful when attempting to reverse-engineer an API.

Verbose error and debug messages

The first category is the now infamous (due to the high instances of information leakage via this method) error category of excessively **verbose error and debug messages**. Application developers include various levels of diagnostic information to aid in the debugging of applications in the field. Users can capture the log and send it to the support team for analysis. Unfortunately, such logging can be overly verbose and, along with useful debug information, can also divulge the specifics of the inner workings of the application and details of the implementation.

As an example, consider the commonly encountered error screen for .NET Core-based web applications shown in *Figure 6.6*:

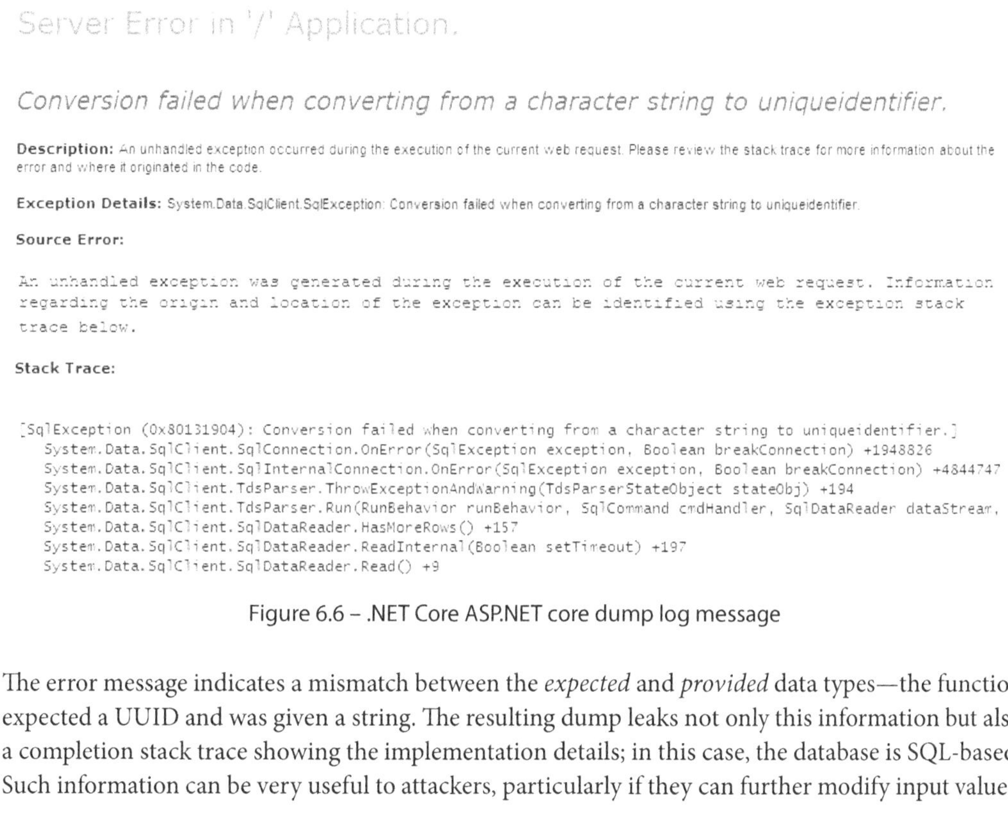

Server Error in '/' Application.

Conversion failed when converting from a character string to uniqueidentifier.

Description: An unhandled exception occurred during the execution of the current web request. Please review the stack trace for more information about the error and where it originated in the code.

Exception Details: System.Data.SqlClient.SqlException: Conversion failed when converting from a character string to uniqueidentifier.

Source Error:

An unhandled exception was generated during the execution of the current web request. Information regarding the origin and location of the exception can be identified using the exception stack trace below.

Stack Trace:

```
[SqlException (0x80131904): Conversion failed when converting from a character string to uniqueidentifier.]
   System.Data.SqlClient.SqlConnection.OnError(SqlException exception, Boolean breakConnection) +1948826
   System.Data.SqlClient.SqlInternalConnection.OnError(SqlException exception, Boolean breakConnection) +4844747
   System.Data.SqlClient.TdsParser.ThrowExceptionAndWarning(TdsParserStateObject stateObj) +194
   System.Data.SqlClient.TdsParser.Run(RunBehavior runBehavior, SqlCommand cmdHandler, SqlDataReader dataStream, Bu
   System.Data.SqlClient.SqlDataReader.HasMoreRows() +157
   System.Data.SqlClient.SqlDataReader.ReadInternal(Boolean setTimeout) +197
   System.Data.SqlClient.SqlDataReader.Read() +9
```

Figure 6.6 – .NET Core ASP.NET core dump log message

The error message indicates a mismatch between the *expected* and *provided* data types—the function expected a UUID and was given a string. The resulting dump leaks not only this information but also a completion stack trace showing the implementation details; in this case, the database is SQL-based. Such information can be very useful to attackers, particularly if they can further modify input values.

The world of APIs is not immune to this type of implementation flaw either, as the sample GraphQL query in *Figure 6.7* shows:

Figure 6.7 – GraphQL query leakage

In this example, an internal fault has led to information leakage that discloses that a local SQLite database is used for storage.

OS and framework enumeration

Returning to the OWASP ZAP scan from previously, a closer look at the alerts panel from *Figure 6.5* reveals two valuable pieces of information detected by ZAP, shown in detail in *Figure 6.8*.

Firstly, the server leaks both the host OS and the web server, as shown in this screenshot:

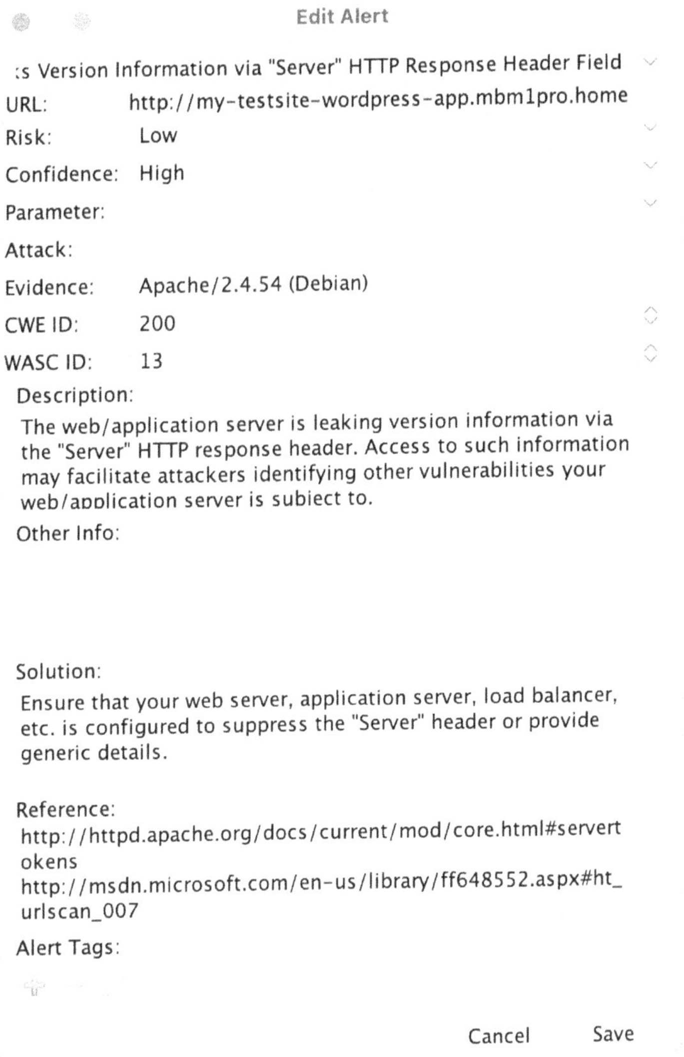

Edit Alert

:s Version Information via "Server" HTTP Response Header Field

URL: http://my-testsite-wordpress-app.mbm1pro.home

Risk: Low

Confidence: High

Parameter:

Attack:

Evidence: Apache/2.4.54 (Debian)

CWE ID: 200

WASC ID: 13

Description:

The web/application server is leaking version information via the "Server" HTTP response header. Access to such information may facilitate attackers identifying other vulnerabilities your web/application server is subject to.

Other Info:

Solution:

Ensure that your web server, application server, load balancer, etc. is configured to suppress the "Server" header or provide generic details.

Reference:

http://httpd.apache.org/docs/current/mod/core.html#servert okens

http://msdn.microsoft.com/en-us/library/ff648552.aspx#ht_ urlscan_007

Alert Tags:

Cancel   Save

Figure 6.8 – Leakage of OS and web server information

The other very useful implementation detail leaked is the fact that the application runtime is PHP 7.4.33, as shown in *Figure 6.9*:

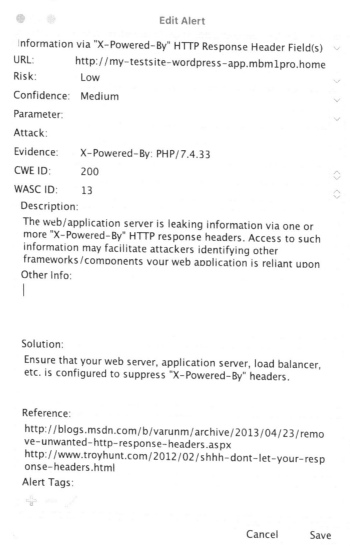

Figure 6.9 – Leakage of application runtime information

This allows an attacker to browse vulnerability databases and identify any exploits known to target this version of PHP.

Timing or volume attacks

Even if an API implementation is well designed to avoid inadvertent leakage of implementation details, a skilled attacker can still gain useful information from an API by examining how the API performs under load and error conditions.

The first type of attack is a **timing attack**, where an attacker monitors the request and response times of the API and then attempts to identify anomalies that reveal details of the underlying implementation. As an example, for a password reset API endpoint, the attacker might observe a `401 Unauthorized` status code with an average execution of 50 ms for some usernames and average execution times of 200 ms. This may seem innocuous to a casual observer; however, what this most likely means is that the usernames taking a longer time to process actually exist in the database and that the longer processing time is the computation of a password match function. This is a quick operation for usernames that do not exist, usually a simple database lookup that fails.

Volume attacks involve bombarding an API with high volumes of data, which causes the API to fail or misoperate and potentially leak data that it should otherwise protect.

Utilizing online tools such as BuiltWith or Wappalyzer

Nowadays, it is almost impossible to provide a service on the internet without one or more third-party services analyzing the traffic, performance, or technology stack used to build the service. Two of the most popular sites at the time of writing are BuiltWith and Wappalyzer, which promise to identify a website's major architectural and structural components. Such details may be useful in the hands of an attacker.

Evading common defenses

In the final section of this chapter, we will examine various ways in which an attacker can evade the defenses of an API-based system.

Regular expression evasion

Regular expressions are one of the fundamental building blocks of many input validation or filter-based defense solutions used to protect APIs. Unfortunately, due to the relative complexity of regular expressions, they are sometimes prone to abuse from an attacker. Whilst it is possible to create a regular expression that achieves the desired result (filters bad input), the expression may also filter good input or be bypassed entirely by an attacker.

As an attacker, we will briefly cover this latter scenario — how do you create attack payloads that can bypass regular expression tests? The first bypass is a brute-force attack that causes the regular expression engine to perform a series of computationally intense operations on the input data and will either crash the engine totally or perhaps bypass the check (depending on how the code is implemented).

This attack exploits two features of regular expressions: greedy matching and backtracking. Consider as an example a regular expression pattern `(x+)*y`, which does the following:

1. Performs a greedy match of the character `x`.

2. Find as many of those occurrences as possible.

3. Find the character `y`.

Now due to the way the regular expression engine works, it will use **backtracking** at the end of the input string and check for a match of the inner pattern `(x+)` — if it gets a match, it will return; if it does not, it will backtrack one character and try again. Unfortunately, the inner regular expression is greedy, so that will attempt to match as much as possible, causing the number of operations required to perform a match to rise exponentially with the length of the input string. By simply passing in a suitable length non-matching string, an attacker can force an impractical number of operations on the engine — even 40 characters will cause the search to take some years to complete! You can try this yourself using the `https://regex101.com/` website to trace the steps.

Whilst a denial-of-service attack is the most extreme form of regular expression abuse, many other attacks can be used to bypass filters based on regular expressions. The best examples of common regular expression bypasses are available in this GitHub repository: `https://github.com/attackercan/regexp-security-cheatsheet`.

Typical techniques involve the following:

- Inserting whitespace characters (spaces and tabs), often as HTML-encoded values

- Using a mix of uppercase and lowercase characters

- Manipulation of carriage return and line feed character patterns

Using a combination of these techniques can exploit poorly implemented regular expression patterns.

For an excellent example of how a poorly formed regular expression can be abused in the real world, please refer to the vulnerability described in *#9: Remote access to two popular vehicles* in *Chapter 4, Investigating Recent Breaches*. In this case, a regular expression to validate input emails accepted a variety of control characters at the end of the email address that should have otherwise been rejected.

Encoding and manipulation

We have already seen how regular expressions can be vulnerable to manipulation of input characters, leading us to the more general topic of encoding bypasses. The concept is quite easily understood — computer systems represent information in several formats (see the discussion on *Encoding* in *Chapter 1,*

What Is API Security? for an overview of the topic). Unfortunately, an attacker can confuse an API by manipulating the format of the input data to bypass input checks (such as the regular expression bypasses) or to inject malicious content into the API backend (a good example is the variety of injection attacks covered in the next chapter, *Chapter 7, Attacking APIs*).

Several different techniques can be used; let us look at the most common.

String termination and whitespaces

Computer systems process binary data; however, much of their data is transferred in a human readable text format. As such, it requires delimiting characters to mark the transitions between words or sentences and termination characters to mark the end of the string or message. Unfortunately, there are a wide variety of options that can be used for either. From an attacker's perspective, there is ample opportunity to confuse the API into accepting malformed data by using unusual combinations of these characters. For example, an attacker may trick an API into receiving multiple lines of input by inserting carriage return characters or may prematurely terminate a string with an unexpected character. Typical real-world examples include log file injection, where an attacker injects whitespace into a log file to try and hide evidence of an attack (by tricking alerting rule logic or simply scrolling messages of the visible screen), and HTTP header injection attacks, where an attacker can inject additional headers to defeat security controls.

As an attacker, use a list of common string termination characters and delimitation characters to try and exploit the API, either by causing data to be malformed or to inject additional data.

Mixing cases

Another construct of the text-based nature of a lot of API data is that of case sensitivity. In many cases, API backend controls and security measures may be written in a case-sensitive manner, and simply by changing the case (switch to all lowercase, or a random mixed case), it might be possible to defeat some of these controls. A typical example is a regular expression used as a blocklist that is case-sensitive, allowing an attacker to use a mixed-case input and bypass this control and exploit the internal system, which may not be case-sensitive.

When attacking using injection attacks (OS command attacks, path traversal, or SQL/NoSQL injection attacks), there is a good likelihood the API will be utilizing a protection mechanism that may be sensitive to a case-switching attack.

Encoding

The final method of tricking an API into accepting data that it would otherwise reject is to use one of the many encoding methods available. Again, the principle is similar: if the API has a protection mechanism that expects data in one format, trick it into accepting data in another format. This technique is typically used to bypass input filters, such as a filter designed to detect script characters such as `<`, `>`, `/`. By encoding the input, an attacker can bypass the filter and attempt to exploit internal systems vulnerable to scripting attacks.

The easiest way for an attacker to experiment with these techniques is to use the BurpSuite Decoder module to convert data from its canonical format into an alternative format. The following example shows how a common **Cross-Site Scripting** (**XSS**) payload can be converted into a Base64 payload:

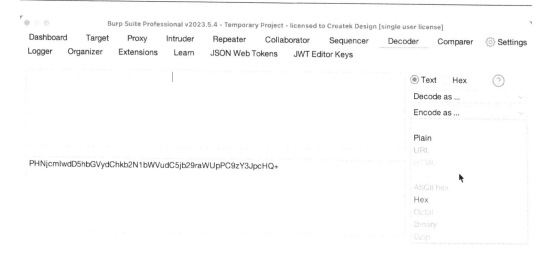

Figure 6.10 – Encoding data in BurpSuite Decoder

When using the BurpSuite Intruder module to attack APIs, it is possible to use the payload processing function to perform a variety of transformations on payload data including modifying the case, adding prefixes and suffixes, matching and replacing, and encoding and decoding into a number of common formats. This feature is shown here:

Clear

Deduplicate

Add payload processing rule

Add   Enter a ne   ⑦   Enter the details of the payload processing rule.

Add from list ...

Select rule type

Select rule type
Add prefix
Add suffix
Payload processing   Match/replace
You can define rules to perfor   Substring
Reverse substring
Add   Enabled   Modify case
Edit   Encode
Remove   Decode
Up   Hash
Down   Add raw payload
Skip if matches regex
Invoke Burp extension
Replace placeholder with base value
Payload encoding   Replace placeholder with collaborator payload

This setting can be used to URL-encode selected characters within the final payload, for safe transmission within HTTP requests.

☑ URL-encode these characters: .\=<>?+&*;:"{}|^`#

Figure 6.12 – Modifying payload data in BurpSuite Intruder

Typically, input filters will block commonly used block control characters used in HTML, CSS, and scripts such as: ( ) < > { } [ ] / \ | ; : ` Start by encoding these using URL or HTML encoding to try and evade any filter that is explicitly filtering these characters.

Another encoding technique that can be used is to switch the character encoding in use by explicitly overriding the API default, typically using the following HTTP header:

```
Content-Type: text/html; charset=UTF-8
```

By changing the character encoding, an attacker may be able to insert unexpected characters into the API backend.

Encoding attacks are something of a black art and will rely on the relative experience of the attacker and the target under attack. The *Further reading* section has excellent references for learning more about this topic.

Rate limiting

One of the most frustrating API defenses for an attacker is rate limiting. Many attacks against APIs will require multiple attempts before succeeding and will be executed repeatedly using scripts or tools such as BurpSuite. Unfortunately for novice attackers, unsophisticated attacks can be easily detected and thwarted by rate limiting in the API gateway or the API implementation.

As an attacker, learning some basic skills to evade rate limiting is essential. The first step is to understand when you have been rate-limited, and this can be determined in a few ways:

- By consulting the documentation for specific rate limits and quotas that are applicable
- By examining common API response headers to indicate the current rate limiting metrics in effect (`RateLimit-Limit:`, `RateLimit-Remaining:`, `RateLimit-Reset:`)
- By receiving a `429 Too Many Requests` status code

Once this occurs, you will need to use one or more of the following techniques to evade the rate limiting.

Slowing down your attacks

The most obvious method of avoiding rate limiting is to slow down the rate of your attacks. If you are fortunate enough to know the applicable rate limits, use the features in your attack tools to stay within these limits. BurpSuite, for example, had many options to insert inter-operation delays and to even use a progressively increasing delay.

Origin header manipulation

If you are lucky, your target API may honor the various *origin* HTTP headers specifying the client and target addresses and IP addresses, and hostnames. Use the features in your attack tool to insert (or rewrite) headers to change the origin or client IP and see whether this defeats the rate limiting

```
X-Originating-IP: 127.0.0.1
X-Forwarded-For: 127.0.0.1
X-Remote-IP: 127.0.0.1
X-Remote-Addr: 127.0.0.1
X-Client-IP: 127.0.0.1
X-Host: 127.0.0.1
X-Forwared-Host: 127.0.0.1
```

Similarly, try and adjust the `User-Agent` header to try and fool the API into thinking it is being accessed by multiple agents. Anything that can confuse the rate limiting logic will likely bypass the limit since the designers would rather err on the side of caution and avoid blocking legitimate traffic accidentally.

IP address manipulation

The final resort for an attacker is to spoof their IP address to defeat the rate limiting. The easiest way to do this is using a proxy server or a **Virtual Private Network** (**VPN**), which offers exit node locations in various global locations. There are numerous low-cost VPN providers that fit this bill perfectly. When selecting a provider, it is worth choosing one that has a **Command-Line Interface** (**CLI**) to allow automation within your attack scripts.

Another option is to use cloud server instances to host a virtual machine to launch your attacks. Most providers will give you an option to refresh your public IP address at will. Be aware that a cost may be associated with using either a virtual machine or a public IP address.

Since, as an API attacker, you will most likely be using Burp Suite, then the IP Rotate (`https://github.com/portswigger/ip-rotate`) plugin is the easiest and most robust method for defeating rate limiting. This plugin uses API gateways on various **Amazon Web Services** (**AWS**) locations globally and re-routes your Burp Suite traffic via different regions in rotation.

This screenshot shows the configuration of the plugin, with all regions enabled.

Figure 6.13 – Using the IP Rotate extension in BurpSuite

Summary

This chapter took us on a complete journey from zero knowledge of a target right through to being able to determine the version of the database used. The passive reconnaissance techniques principally used Google and the query operators, as well as the Shodan database, to determine a range of likely candidate targets for exploration.

We learned how to use active reconnaissance to actively focus on the API implementations on the hosts by examining their behavior under live probing using nmap or Massscan, while OWASP ZAP can provide a wealth of insight using spider scanning. Finally, we learned how to use information leakage to gain insight into the inner details of the implementation, allowing us to understand details such as the host OS and database.

By conducting a thorough discovery phase, you will have placed yourself in a perfect position to move on to the final chapter in the section—on attacking APIs.

Further reading

Further reading on discovering APIs:

- `https://nordicapis.com/13-api-directories-to-help-you-discover-apis/`

Further reading on Google operators and the GHDD:

- `https://www.googleguide.com/advanced_operators_reference.html`
- `https://www.exploit-db.com/google-hacking-database`
- `https://github.com/jakejarvis/awesome-shodan-queries`

Further reading on nmap:

- `https://nmap.org/book/toc.html`
- `https://owasp.org/www-pdf-archive/Analysing_Networks_with_NMAP.pdf`

Further reading on the interception of data on devices:

- `https://approov.io/blog/how-certificate-pinning-helps-thwart-mobile-mitm-attacks`
- `https://learning.postman.com/docs/sending-requests/capturing-request-data/capturing-http-requests/`

Further reading on excessive information exposure:

- `https://owasp.org/www-community/Improper_Error_Handling`
- `https://docs.42crunch.com/latest/content/extras/protection_security_headers.htm`

Further reading on evading common defenses:

- `https://cheatsheetseries.owasp.org/cheatsheets/XSS_Filter_Evasion_Cheat_Sheet.html`
- `https://s0md3v.medium.com/exploiting-regular-expressions-2192dbbd6936`
- `https://github.com/attackercan/regexp-security-cheatsheet`
- `https://portswigger.net/web-security/essential-skills/obfuscating-attacks-using-encodings`
- `https://github.com/portswigger/ip-rotate`

7
Attacking APIs

In the previous chapter, we explored the passive and active techniques that can be used to discover APIs. The focus in this chapter moves on to actively attacking and exploiting those APIs using various methods to exploit vulnerabilities in the API design or implementation. By the end of this key chapter of the book, you will be able to attack APIs on your own using various techniques. As a builder of APIs, one of the best ways to test their defense is to attack them yourself.

Secure APIs rely on strong authentication and authorization. In the first topic, we will learn how to attack by identifying design and implementation weaknesses. Fuzzing and brute force attacks are among the easiest to perform, and you will learn how to use automated attacks to *crack open* an API. Next, the focus turns to data-based attacks: either learning how to trick an API into accepting more data than expected (often leading to injection-based attacks) or finding APIs that leak excessive information.

The chapter concludes with a look at some common techniques that can be leveraged to exploit the business logic of an API since, often, these vulnerabilities can be the hardest to defend against. Finally, we will examine some techniques to evade common API defenses.

In a nutshell, this chapter will cover the following main topics:

- Authentication attacks
- Authorization attacks
- Data attacks
- Injection attacks
- Other API attacks

Let us begin our exploration by examining the most attractive attack vector when attacking an API: authentication attacks.

Technical requirements

For this chapter, you will need a development machine capable of the following:

- • Running Docker locally

- • Running VS Code with various marketplace extensions

- • Access to the internet and a GitHub account to access the examples

This chapter contains many code samples in various languages; these can either be run locally, which will require the installation of compilers, SDKs, and frameworks, or from within a Docker build container.

The example code and various breaking changes to the instructions can be found in the `Chapter 7` folder on the book's GitHub repository here: `https://github.com/PacktPublishing/Defending-APIs/tree/main/Chapter7`

Authentication attacks

APIs are secured by controlling access based on a client's identity (their authentication) and their permissions (their authorization). One of the most obvious and popular means to attack an API is to bypass the authentication controls by impersonating a client. Typically, this is done by guessing access credentials, stealing or forging credentials, or exploiting weaknesses in the authentication logic.

Insecure implementation logic

From an attacker's perspective, there are two primary attack vectors: attacking a design weakness or exploiting an aspect of insecure implementation logic. Let us look at them in the following sections.

Credential attacks

For an API with a human user (where the authenticated user is a human or is delegated access to an authorized intermediary, such as an OAuth2 client), there will almost always be the need for credentials to be provided by the user. This is one of the most obvious points for an attacker to gain access from using one of the following methods.

Credential stuffing

The technique of **credential stuffing** involves the use of credentials (username and password combinations) harvested on the internet (typically on the so-called **darknet** or similar black market hacker sites) to attempt to access a system. This technique is reliant on the human behavior of re-using passwords across multiple websites rather than using unique passwords per site. Unfortunately, if credentials are leaked via a breach on one site, then they will invariably appear for sale on black market sites.

The barrier to entry for an attacker is relatively low since these lists are becoming more readily available and are increasing in size as more breaches occur over time. A successful credential-stuffing attack typically requires a very large word list. Attackers will typically use a distributed network or a botnet to launch a credential-stuffing attack to avoid triggering intrusion detection mechanisms on target sites.

From a defender's viewpoint, credential-stuffing attacks can be mitigated using the following methods:

- Use **multi-factor authentication** (**MFA**), particularly as part of a **step-up authentication** process (where higher-order authentication, such as MFA, is required for more sensitive operations or new devices)
- Monitor email addresses for indication of compromise — various third-party services can monitor whether an email has appeared in high-profile breaches (for example, `https://haveibeenpwned.com/`)
- Actively monitor for bot access since bots are often used for credential-stuffing attacks
- Monitor IP address ranges for reputation scoring since credential-stuffing attacks often originate in so-called bot farms or server farms, which have well-known IP address ranges

Credential stuffing is, in most cases, likely to be the most successful of the credential attacks. As an attacker, try to find good resources for username and password lists and to keep these updated.

Password spraying

While a credential stuffing attack uses harvested credentials leaked from incidents, a **password spraying** attack differs by trying to infer valid credentials by utilizing common sets of usernames and passwords to attack a login endpoint (see the discussion in the *Fuzzing API endpoints* section of *Chapter 5, Foundations of Attacking APIs* on fuzzing tools and techniques). A password spraying attack is similar to a dictionary attack (where a wordlist is used as the source credentials); however, the number of passwords is limited to a smaller set and the number of users is maximized. The intent behind a spraying attack is to minimize the likelihood of triggering the protection mechanisms designed to prevent brute force attacks.

A password spraying attack cycles through a large list of users and tries a few passwords at a time per user. By only making a few attempts per user, the chances of locking out a user account are minimized. To set up a password spraying attack, the attacker should perform some reconnaissance of the target site to identify both the maximum number of failed attempts for password submission and the timeout period of the protection window for failed attempts.

Using a tool such as Burp Suite's **Intruder** feature, an attacker can easily construct a password spraying attack. Consider the example of an API providing a /login method taking a username and password:

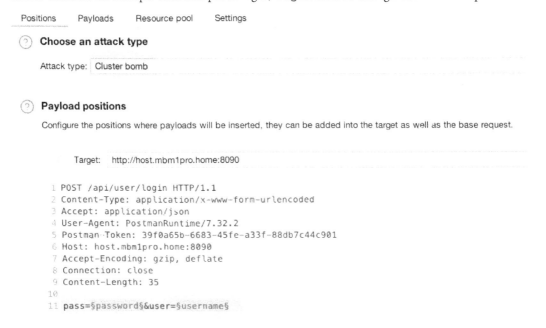

Positions   Payloads   Resource pool   Settings

(?) **Choose an attack type**

Attack type: Cluster bomb

(?) **Payload positions**

Configure the positions where payloads will be inserted, they can be added into the target as well as the base request.

Target: http://host.mbm1pro.home:8090

```
 1 POST /api/user/login HTTP/1.1
 2 Content-Type: application/x-www-form-urlencoded
 3 Accept: application/json
 4 User-Agent: PostmanRuntime/7.32.2
 5 Postman-Token: 39f0a65b-6683-45fe-a33f-88db7c44c901
 6 Host: host.mbm1pro.home:8090
 7 Accept-Encoding: gzip, deflate
 8 Connection: close
 9 Content-Length: 35
10
11 pass=§password§&user=§username§
```

Figure 7.1 – Burp Suite Intruder attacking an API login method

In this example, the method takes a password and username parameter in the body. By replacing the values with parameters (§password§ and §username§), the Intruder feature can replace these with values from a configurable payload.

First, set up the first payload with the set of passwords to be used as such:

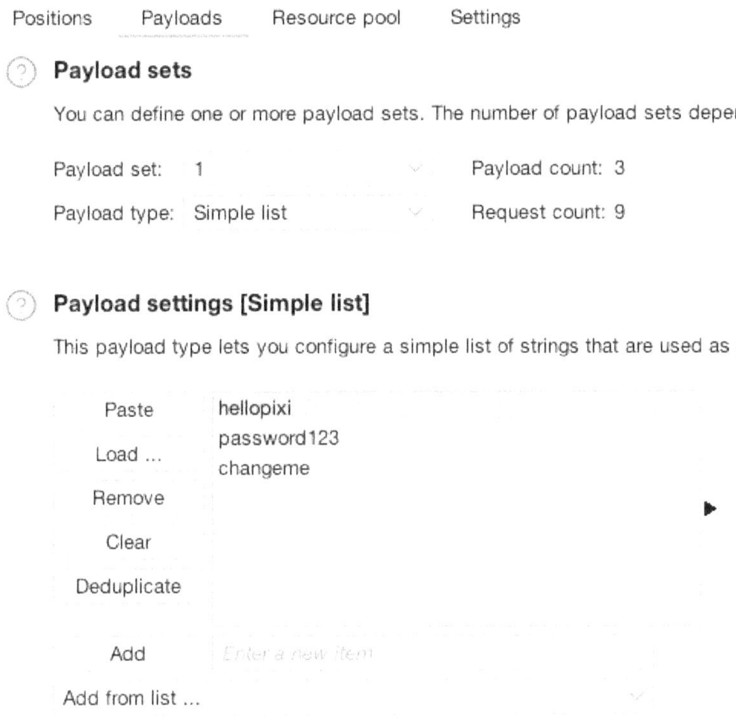

Positions   Payloads   Resource pool   Settings

② **Payload sets**

You can define one or more payload sets. The number of payload sets depends on the

Payload set: 1   Payload count: 3

Payload type: Simple list   Request count: 9

② **Payload settings [Simple list]**

This payload type lets you configure a simple list of strings that are used as payloads.

Paste	hellopixi
Load ...	password123
	changeme
Remove	▶
Clear	
Deduplicate	
Add	_Enter a new item_
Add from list ...	

Figure 7.2 – Intruder password payload

Second, set up the second payload with the set of usernames to be used like so:

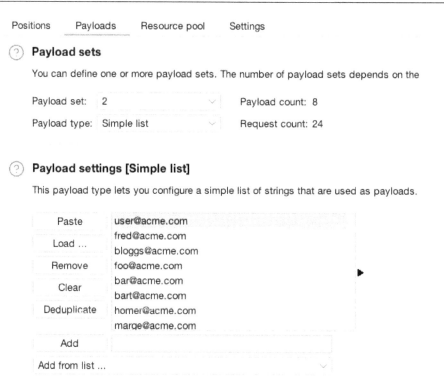

Figure 7.3 – Intruder username payload

In constructing these payloads, I used a large username payload with a small set of passwords. The final step is to execute the intrusion attack in **cluster bomb** mode (meaning every permutation of username and password combinations is attempted), which gives the following results:

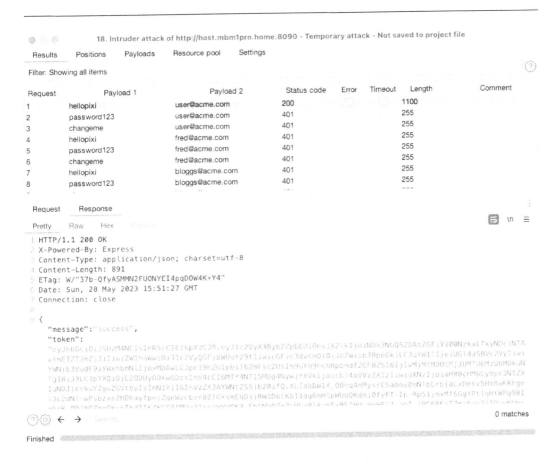

Figure 7.4 – Cluster bomb attack results

Sorting the results by status code reveals the username and password combination that worked. The response body includes an access token for future transactions.

The Intruder feature can be finely configured to set an interval between requests (including adding random intervals and increasing delays) to prevent triggering brute force protection. A sample configuration is shown in the following screenshot:

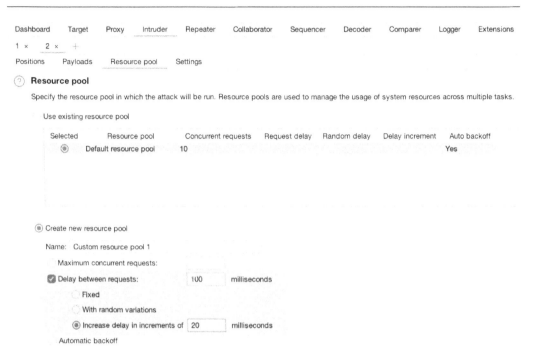

Figure 7.5 – Configuring Intruder with delays

There is also an extension for Burp Suite called **Burp Password Spray**, which can aid with introducing additional delays to accommodate lockout reset times.

Brute forcing

The final variant of a credential attack is the least subtle of the methods: the **brute force attack**. This method uses (very) large username and password wordlists and attempts to brute force the login function using every possible combination of values. Again, the Burp Suite Intruder tool can be used in cluster bomb mode, as can other specialized tools, such as **wfuzz** or **Gobuster**.

Key-based attacks

Keys are similar in functionality to tokens (refer to the callout at the end of this section for further information), as they represent a client's identity. In the simplest terms, they can be considered a replacement for a username and password combination. Since they are often longer lived than tokens and provide much broader access, they present a good attack vector for the aspirant API hacker.

Hardcoded keys

One of the easiest attacks to make against keys involves finding a hardcoded key in the application configuration or the application binary. The application might be configured with a hardcoded development environment key during the development lifecycle for expediency. Unfortunately, the application may be accidentally shipped with a hardcoded production key. The best practice would be to ensure that the keys are distributed and managed securely using a key vault; however, as is the nature of software development, this practice is not always rigidly enforced. *Chapter 4, Investigating Recent Breaches,* contains an excellent example of how hardcoded keys can compromise a mobile application in the *Microbrewery application* section. In this case, the application has been shipped with a hardcoded key in the binary that could easily be retrieved using simple debugging tools.

Leaked keys

Hardcoded keys typically imply that keys have been committed to a source code repository, and this anti-pattern leads to the second common attack vector, namely leaked keys. If keys are committed to a repository (or other documentation sources), it becomes increasingly likely that they will become accessible if the access control to the repository is not sufficiently controlled. Unfortunately, it is common for source control or documentation systems to be left with completely open access to allow access to the keys. Keys might also be leaked by third-party systems, such as CI/CD platforms, or deliberately stolen and leaked by disgruntled staff or other parties.

For an attacker, leaked keys are one of the best attack vectors to access an API. *Chapter 5, Foundations of Attacking APIs,* covers the tools and technologies used for discovering API keys in the *Finding API keys* section.

Weak keys

In the same way that a weak password can be attacked using brute force methods, weak keys can be vulnerable to similar brute forcing or cracking attacks. Since passwords will usually need to be remembered by a user, they are (all too) frequently shorter and less complex than would be desirable for optimum security. While keys are usually stored on a client and do not need to be remembered, they often suffer from weaknesses in their generation that allow them to be cracked. Such weaknesses include being too short (keys of less than twelve digits are still encountered), having low entropy (a lack of randomness), or, in some cases, being predictable.

Once you have found an API endpoint accepting a key (or token), you can use a combination of tools to attempt to brute force the key. The first step is to determine whether the key is weak and susceptible to brute force. For this, the **Burp Suite Sequencer** (`https://portswigger.net/burp/documentation/desktop/tools/sequencer`) tool is ideal. This will analyze a sample of keys (100 or more) and perform a mathematical analysis on them to test their quality regarding dimensions such as randomness, entropy, and correlation. As an illustration, I captured the results from an API's /register function, which returned a **JSON Web Token** (**JWT**) with a few identity claims. These tokens were then loaded into Sequencer, which produced the summary analysis shown in the following screenshot:

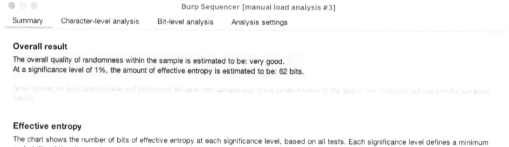

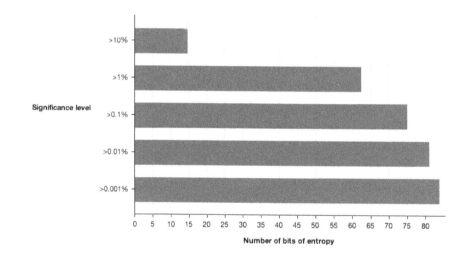

Figure 7.6 – Analyzing token randomness in Burp Suite Sequencer

As noted, the overall quality of randomness is very good, meaning that these tokens would not be good candidates for a brute force attack. If the sequencer can identify any patterns, repeated sequences, and so on, then the character-level analysis can suggest good locations to attempt to attack using a brute force tool, such as Burp Suite Intruder or Wfuzz (covered in *Chapter 5*, *Foundations of Attacking APIs* in the *Fuzzing API endpoints* section).

Lack of key rotation

Although keys are designed to be a disposable item (easy to create, use, and destroy), it is often the last part that is the most troublesome – keys are often persisted into binaries, or hard-coded, making it difficult to rotate them in the case of a compromise. As an attacker, always be on the lookout for keys that have not been revoked. In *Chapter 5*, *Foundations of Attacking APIs*, the *Finding API keys* section demonstrated various techniques for harvesting API keys from public repositories. These can either be tested directly against target API keys for validity or using a tool such as **Trufflehog** (`https://github.com/trufflesecurity/trufflehog`), which can verify whether the keys are still valid.

> **Keys versus tokens**
>
> The terms *keys* and *tokens* are used somewhat interchangeably with reference to API authentication and authorization. Although they perform a similar purpose, some subtle differences are worth understanding.
>
> A *key* is usually a simple replacement for a username and password combination and simply identifies the client to the server and nothing more. Keys are usually used when the client is accessing a service on a server rather than any data, records, or resources. As an example, a user of the *Giphy* GIF-sharing service is identified by a key, which is how the *Giphy* service grants access to their service. The user does not have any resources on the *Giphy* server.
>
> A *token* offers the same features as a key (proving identity), but it typically contains additional *claims* (individual pieces of information about the token holder), such as identity, permissions, scopes, and expiration times. When attempting to access a server resource (for example, a document), the server will assess the claims in the token against the permissions granted to that user and allow or deny access accordingly.
>
> Keys can be considered more lightweight than tokens (as they are designed to be easily issued and disposed of and do not have an expiration date), while a token usually contains more data and should be protected with secure generation and distribution schemes (such as OAuth2).
>
> For our discussion in this chapter, keys and tokens can be attacked using similar methods unless otherwise specified (such as JWT-specific attacks).

Token-based attacks

Since JWTs are one of the fundamental building blocks of APIs, it stands to reason that they are an obvious attack point for an aspiring API attacker. Due to the relative complexity of generating and safely using JWTs, they are frequently implemented insecurely and easily exploited. This section will examine some of the most common JWT attack techniques. If you need a refresher on the fundamentals of JWTs, please refer to *Chapter 2*, *Understanding APIs*, and the section on *Using JSON Web Tokens for claims and identity*.

Before starting any attacks, you will need to capture and inspect JWTs. In Burp Suite, this is best accomplished using one of the marketplace extensions, as shown in the following screenshot:

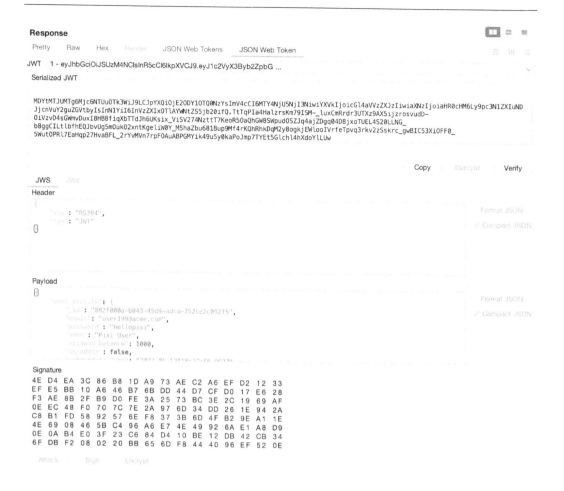

Figure 7.7 – Inspecting JWTs in Burp Suite

Armed with this ability to intercept and modify JWTs from the target API, let us understand the different attacks, from the easiest to the hardest.

Missing verification

As an attacker, if you are lucky, you will discover an API implementation that fails to verify the integrity of the JWT. The correct sequence for consuming a received JWT is to verify the signature first and then to decode the payload — only after both operations are complete can the content be trusted. Unfortunately, developers will frequently fail to perform the verification step, meaning they blindly trust the JWT payload (remember, the payload can be modified in transit and is protected by a signature). This is in some part due to confusion over JWT client libraries, which sometimes provide a `decode()` method that does verification, while other times, a specific `verify()` method must be invoked.

As an attacker, it is worth simply modifying a JWT within your attack tool and seeing whether the API still accepts it. If it does, you have an API that is failing to verify the signature.

The None attack

Related to the missing verification case is the so-called None attack. In this scenario, the developer will have deliberately chosen to omit the signature stage of the JWT creation process by specifying the None parameter as the signature algorithm. The option for no algorithm is specifically intended for cases where the integrity of the token can be validated by some other means (externally signed, for example). Unfortunately, providing this option at all has meant there are numerous instances where `alg=None` has been encountered; there is even a website dedicated to tracking occurrences encountered in the real world, found at `https://www.howmanydayssinceajwtalgnonevuln.com/`.

This token is easy to spot by inspection or after decoding, as shown here:

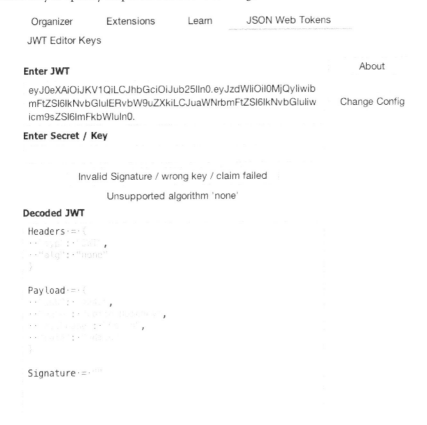

Figure 7.8 – The None algorithm

First, notice in the payload itself that the third section (the signature) has been omitted; for a signed JWT, a signature would be present after the second period. It is trivial to spot this type of token on the wire (in Burp Suite or using a proxy tool).

Second, once the token is decoded, the `"alg"`: `"none"` field clearly indicates that the JWT is unsigned, as is the blank signature field. Note also the warning from Burp Suite stating that it could not validate the signature. If, as an attacker, you find such a token, you are free to modify the contents and resubmit it to your target, perhaps after you have elevated your privileges to an admin role, as shown.

The algorithm switch attack

The algorithm switch attack exploits some common misunderstandings of symmetric versus asymmetric cryptography (for a refresher, please see the *Cryptography* section of *Chapter 1, What Is API Security?*) and/or flawed implementation of JWT libraries. For a symmetric algorithm, the same key is used for signing and verification. For an asymmetric algorithm, a secret key is used for signing and the corresponding public key is used for verification.

This attack results from confusion in the verification method in the client library, which can be tricked into using a public key (used for asymmetric algorithms, such as RS256) as the verification key for a symmetric algorithm, such as HS256. Since the public key is public, an attacker can create a malicious token signed with this key and then trick the client into using this same key in the verification process.

JWT header parameter injections

A JWT typically specifies an `alg` and `typ` value in the header; however, several other fields can be used to specify parameters, most notably keys and keysets. For example, the `jwk` specifies a JSON object as a key and a `jku` provides a URL from which keys can be obtained. Since these key values or references are externally specified, there is an opportunity for an attacker to modify these keys using techniques such as directory path traversal or key substitution attacks using Burp Suite.

For example, the following snippet shows a key ID (`kid`) referencing a local file path, which could be subject to manipulation to point to self-signed keys:

```
{
    "kid": "../../my/bad/path",
    "typ": "JWT",
    "alg": "HS256",
    "k": " NazJZBuNhQ5DBHW0APemqQ"
}
```

This is a complex attack type, and the reader is advised to review the excellent resources on this topic from PortSwigger (`https://portswigger.net/web-security/jwt`).

Cracking JWTs

The final attack on JWTs that we will examine is the cracking of the key itself using brute forcing techniques. For asymmetric algorithms (such as RS256), the keys are typically large (1,024 bits or greater) and not susceptible to this method. However, for symmetric algorithms (such as HS256), developers will sometimes use passphrases (which are easily memorable) instead of higher entropy (longer) keys. Exactly the same techniques used to crack passwords can then be applied to crack the symmetric keys.

For example, the `jwt_tool` (covered in *Chapter 5*) can apply a wordlist (comprising potential keys) against a JWT to identify the signature key. Once the key has been identified, the attacker can create new forged JWTs at will since they have the signing key. When it comes to a wordlist-based attack, the real challenge is finding a good wordlist, as many online lists are outdated.

Although it is important to be aware of this attack vector, the likelihood of success with this method is diminishingly small since even a moderate key length will require an unrealistic amount of time to crack (even eight or nine characters could take hundreds of days of raw brute forcing).

Attacking design weaknesses

Finally, let us focus on a few common design weaknesses encountered in APIs relating to authentication.

Weak reset process

We have already described the challenges with the rotation of API keys and how they allow attackers to compromise an API. Related to the rotation of keys is doing a password reset whenever a user interacts with an API-based platform. Rather than attacking the API itself, an attacker may choose to attack the user and gain access by compromising their account(s).

As an attacker, the first step is identifying how an end user would initiate a password reset. Typically, this involves a reset sequence being initiated for the user's email. A new password or a reset link is usually sent to the user. Using a test account, you can examine the details of this process to identify weaknesses. In the case of a new password, does the platform enforce a password reset on the next login? Does the password appear predictable at all?

In the case of a reset link, this will take the form of a URL embedded in the email, similar to this example:

```
https://www.acmecorp.com/forgot_password.php?user_id=0001&token=66001
```

There are a few obvious attack vectors here. First, can we submit other user IDs and potentially take over other users' accounts? Second, is the token easily guessable and open to abuse?

Sometimes, the reset sequence will be initiated by sending a PIN code to a mobile device via SMS or voice call. Again by intercepting the token (on a test account), the attacker can determine whether the tokens are predictable, if they can be re-used, if they expire, and if rate-limiting restrictions are applied.

For further reading, the case study in *#10: Smart scale* in *Chapter 4, Investigating Recent Breaches* demonstrates several account reset design flaws, including arbitrary account takeovers, predictable PINs, and failure to invalidate a reset sequence.

Insecure transmission

One of the quick wins for an API attacker is discovering information transmitted insecurely between a client and an API. There are a number of common locations to check for implementation weaknesses:

- **Insecure Transport:** Although encrypted transport via **Transport Layer Security** (**TLS**) is almost ubiquitous, an attacker should always check if an insecure channel exists to the same resource, i.e., they should try and use HTTP instead of HTTPS.

- **Transmission of sensitive data in URLs:** One of the most common API implementation weaknesses is the transmission of sensitive data in URLs, particularly in the case of GET requests. Even if TLS is used, any parameters will be visible to a machine in the middle and may be persisted in log files. If sensitive data is required for a read operation (usually a GET operation), then the endpoint should be implemented as a POST operation, passing sensitive values in the body.

- **Weak encryption mechanisms:** Always be on the lookout for data that appears to be encrypted and investigate whether it has been encrypted with a weak algorithm, a key that can be cracked, or a key that is exposed.

- Transport security can easily be tested for weaknesses by an attacker using third-party tools such as SSL Server Test, found at `https://www.ssllabs.com/ssltest/`.

Lack of rate limiting

One of the simplest ways an attacker can attempt to subvert an API is by brute force, whether this is attempting to guess credentials, crack encryption keys, or conduct a denial-of-service attack. Although technically simple in concept, rate-limiting implementations often leave much to be desired, and an attacker should always consider attacks using brute force. The final section of this chapter covers a few techniques an attacker can use to bypass rate-limiting controls.

Understand common terminology

This chapter uses some specific terms that may not be familiar to some readers. Here are explanations of common terms:

- **Side-channel attacks**: A side channel is considered an attack against the underlying implementation of a system rather than directly against the system itself. This attack relies on detailed knowledge of the system to know which implementation details to attack. For example, it is possible to retrieve data from semiconductor memory systems after they have been erased, or information can be read from hard drives by analyzing the activity of the drive status LED.

- **Timing attacks**: A timing attack is an excellent example of a side-channel attack whereby an attacker carefully monitors the time taken to execute API commands under various attack conditions. For example, a longer execution time may indicate a successful username lookup and password verification. A shorter time likely indicates that the lookup was unsuccessful and that the user does not exist at all.

- **A-B(-A) testing**: A-B testing is typically used for testing API authorization attacks. First, create a resource as user A. Then, attempt to access the resource as user B. If you can, then you have an authorization issue. As user B, attempt to modify or delete the resource and attempt to access it again as user A. This is an A-B-A test.

- **Reflected vs. persisted attacks**: This attack type is commonly associated with **cross-site scripting (XSS)** attacks. If the attack payload is written into a database permanently, then it is called a persisted attack. If the attack only exists in the current session, then it is a reflected attack. By reloading the session or browser, the attack will be cleared. A persisted attack, on the other hand, is permanently written to the database.

- **Blind attacks**: A blind attack is when an attacker is unable to directly observe the results of his attack. Examples include deleting database tables or user data without receiving a response (because there might not be a terminal session or output window visible to the attacker).

- **Volume attacks**: A volumetric attack is an attack where abnormally large volumes of data are submitted against the target to cause it to misoperate. An example is a **denial-of-service (DoS)** attack against an API.

Authorization attacks

We now have a thorough understanding of the wide array of attacks that can be used on the authentication mechanisms of an API. Let us now turn the focus to the counterpart of authentication: authorization.

Object-level authorization

As a reminder, **broken object-level authorization (BOLA)** occurs when an API becomes confused about the right to access an object and allows unauthorized access. In *Chapter 3, Understanding Common API Vulnerabilities*, this is covered in detail in the *API1:2019 — Broken object-level authorization* section.

Conceptually speaking, BOLA attacks are simple to originate using the following recipe:

1. Identify an API operation that takes an *object ID* as a parameter.

2. Create a resource for the first user (call them `user A`).

3. Confirm that `user A` can access the new resource.

4. Using a second user (call them `user B`) who does not have access to the new resource, attempt to access the same resource. If you succeed, you have identified an API operation susceptible to BOLA.

5. The first challenge with launching BOLA attacks is identifying the object IDs in a request. In some cases, this will be almost self-evident (a numerical value as the last parameter in a request), though other times it's more complex (nested objects within a request body). Fortunately, using tools such as Burp Suite makes it a simple process to identify potential object IDs on endpoints. Some common examples of object IDs are as follows:

    -   Integers as the final parameter in a GET request or in a POST request body

    -   An email address as the final parameter in a GET request or in a POST request body

    -   A UUID as the final parameter in a GET request or in a POST request body

    -   Predictable keys or tokens in GET requests or in a POST request body

    -   Once the object IDs are identified, the attack can be launched using Burp Suite Intruder or Repeater to identify endpoints.

Function-level authorization

As a reminder, **broken function-level authorization (BFLA)** occurs when an API becomes confused about the right to access a function (method) and allows unauthorized access. In *Chapter 3, Understanding Common API Vulnerabilities*, this is covered in detail in the *API5:2019 — Broken function-level authorization* section.

The vulnerability occurs when an API allows an unauthorized user to perform an action on an API function. A few common examples are as follows:

-   Deleting another user's data (access to the DELETE method is not properly protected)

-   Forcing an elevation of privileges on a function by parameter or request manipulation (for example, adding an `admin=true` parameter)

-   Accessing hidden administrative functions (such as an `/admin` endpoint) and executing privileged operations

- The attack methods for broken function-level authorization are similar to those used to detect broken object-level authorization, namely the A-B-A method:

  I.   Create a resource as user A.

  II.   Attempt to modify the same resource as user B (who does not have access to this resource).

  III. As user A, determine whether the resource has been changed. If so, you have an instance of broken function-level authorization.

  IV.   Based on my experience of analyzing real-world API vulnerabilities, the most common broken function-level authorization attacks come from attacks on hidden administrative functions. For example, many content management systems host an administrative endpoint on the /admin or /administrator URL. As an attacker, use a list of well-known administrative endpoints to identify ones that may be unprotected.

Data attacks

Data attacks (covered in *Chapter 3, Understanding Common API Vulnerabilities*) are one of the most common attack vectors for an attacker since, inevitably, the data is most valuable on the black market. While data attacks are most commonly *read* attacks (mapping to the *API3:2019 — Excessive data exposure category* section), the *write* attack (mapping to *API6:2019 — Mass assignment*) can be damaging to an API since this can be used to modify permissions and confidential data of users. Let us look at both types of data attacks.

Excessive information exposure

The numerous API breach examples covered in *Chapter 4, Investigating Recent Breaches*, have one thing in common: they all involve excessive information exposure. APIs can expose excessive information in several ways, and as an attacker, you need to know where to look for it.

The most obvious place is in the response itself — always make sure you are looking at the API responses either in your API client (curl or Postman), in-browser tools, or via a reverse proxy (I almost always use a proxy such as Charles or Burp Suite to capture requests and responses as I explore an API). Sometimes the data may not be obvious in a test view, so make sure to view responses as JSON and XML and be aware that data may also be encoded in various schema, such as Base64 (use Burp Suite to decode this on the fly).

A top tip for discovering an API's data hierarchy is to observe the API and the responses while using the associated application to exercise the API. Use a machine-in-the-middle attack against a mobile application with a reverse proxy to build up the data hierarchy and understand what endpoints handle which data types. Once you have this overview, you can launch more targeted attacks to exfiltrate specific data.

Another way to exfiltrate data is to abuse the API's pagination schemes. If ever you encounter an API request of the following form, you have a candidate for abuse:

```
GET /api/posts?offset=0&limit=10
```

The extra parameters allow the client to specify which starting index (record count) to use and how much data to return per request. Using the Intercept or Repeater tools in Burp Suite, executing a sequence of requests against this endpoint is trivial to extract all the underlying data.

If you are really lucky, you might find a request parameter controlling the amount of detail or verbosity of data returned, as shown here:

```
GET /api/posts?verbose=true
```

It is worth adding these parameters to other requests and checking whether they provide more verbose responses.

Other useful locations to examine for data leakage include logging endpoints (CI/CD systems frequently have well-known endpoints where interesting log files can be retrieved, often containing tokens and keys) and user profile or information endpoints. Always examine the full details returned by an API when viewing your own profile page, as the API may return more than is shown in the client application.

Mass assignment

Mass assignments are a relatively infrequently encountered attack vector, but they can be extremely useful for an attacker to modify server-side data. In particular, this attack is useful to attackers trying to modify permissions and roles (to elevate privilege) or to manipulate pricing, stock levels, or order information.

To understand the first scenario, let us examine the data submitted when a user registers on our vulnerable API:

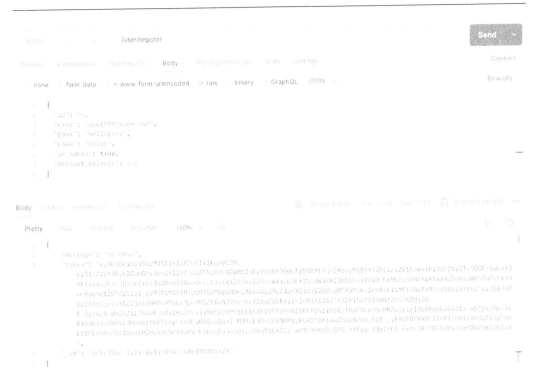

Figure 7.9 – Pixi user registration

In this example, the user is created with an administrator role and an account balance of 100. If an attacker could observe the registration request, they would be aware of the fields called is_admin and account_balance and could attempt to manipulate these on their own profile.

The attacker accesses the /edit_info endpoints, sets their access level to administrator, and increases their account balance to 1000. Unfortunately, this endpoint is susceptible to mass assignment, and a 200 OK result is returned — the attacker has been able to modify the internal details of their account, as shown in the following screenshot:

Figure 7.10 – Successful mass assignment attack

In many cases, the attacker does not have knowledge of the variable names and so must resort to educated guesswork. The first option is to use a good wordlist of API parameter names and try these manually or via Burp Suite. A better option is to use the **Arjun** tool (`https://github.com/s0md3v/Arjun`), which can automate parameter fuzzing either as a parameter or within request bodies. The tool has various options for import and export and fine-grained controls to prevent the triggering of rate limit protections.

Failing either of these options, the attacker should attempt to determine the technology stack used in the API since this may give them clues about the various built-in variables. With this knowledge, an attacker can launch attacks using these variable names as parameters and hope they are processed by the backend. For example, in a .NET Core context, the following code snippet illustrates a typical authorization control on an endpoint:

```
[Authorize(Roles = "Administrator, PowerUser")]
public class ControlAllPanelController : Controller
{
  public IActionResult SetTime() =>
    Content("Administrator || PowerUser");
    [Authorize(Roles = "Administrator")]

  public IActionResult ShutDown() =>
    Content("Administrator only");
}
```

If you know you are attacking a .NET Core API, attempt to inject a parameter named `Roles` with an `Administrator` value and see if you gain elevated access.

Injection attack

Injection attacks are one of the oldest and most well-known software vulnerabilities. Although originally a common defect in web applications, they are quite frequently encountered in APIs, particularly when an API accepts user-supplied input. We will cover the basics of the main injection attack types. Readers wanting more details on topics including hands-on laboratories are advised to take a look at the *Further reading* section in this chapter, particularly the PortSwigger resources.

Detecting injection vulnerabilities

The concept of an injection attack is extremely simple: where an API expects input data, inject one of several types of payloads and observe the behavior. For example, by passing JavaScript into a website form, can you force the website to display a message box? If so, you have found a webpage that is vulnerable to XSS attacks.

For APIs, there are numerous locations where inject payloads can be placed:

- Query strings in the URL (for example, a file path that can be manipulated)
- Tokens (we examined key path injections in JWTs earlier in the chapter)
- Parameters in `POST`/`PUT` requests (we will see an example of a NoSQL injection in a password body)
- Headers
- Detecting your injection attempt's success can be a fine art. In many cases, the end result will be to crash the API totally or return a generic failure method. Always examine the error messages returned very closely, as these will give you a big clue about the next steps. For example, in the case of an attempted SQL injection, an error message including `bad SQL syntax` indicates that you have found an endpoint potentially vulnerable to SQL injection. Attempt to tweak the query further until you succeed. Similarly, an attempt at path traversal with an error message including `path not found` likely means you have succeeded at a traversal but to a non-existing path. Repeat the attack with other paths until you succeed.
- Despite the maturity of this vulnerability type and the wealth of guidance around mitigation or avoidance, injection attacks are still prevalent, and many excellent guides and tools exist to exploit them.

SQL injection

First up, let us look at the best known of the injection attacks: the ubiquitous SQL injection. This is most easily explained with an example (taken from the PortSwigger Academy website at `https://portswigger.net/web-security`). Imagine a URL taking a parameter called `category`:

```
https://insecure-website.com/products?category=Gifts
```

In a backend implemented in SQL, this will typically result in a query to the database using the following syntax:

```
SELECT * FROM products WHERE category = 'Gifts'
```

Now, this is if an attacker were to construct a URL with a well-known SQL injection payload:

```
https://insecure-website.com/products?category=Gifts'+OR+1=1--
```

This would modify the underlying SQL query to the following:

```
SELECT * FROM products WHERE category = 'Gifts' OR 1=1--'
```

This query instructs SQL to return all records where the `category` is equal to `Gifts` or where `1` equals `1` (this is all the records because this is always true and the operation is logical OR-ing the arguments).

This basic SQL injection attack illustrates how simple yet powerful they can be. The trick is normally to know what SQL injection payload to use; fortunately (for an attacker), there are several well-known patterns to attack a vulnerable endpoint. The most basic form of attack (as we have seen) can perform basic operations using query manipulation. However, sophisticated attackers can use UNION attacks to query data from other tables, change application logic, and examine the structure of the underlying database.

There are also excellent attack tools an attacker can use to exploit SQL injection vulnerabilities. The most well-known of these is the **sqlmap** (`https://sqlmap.org/`), which allows an attacker to automate SQL injection attacks against a target and can be used in conjunction with Burp Suite.

NoSQL injection

Database technologies have evolved over the last two decades, and the family of NoSQL databases (of which MongoDB is the most well known) has emerged as a natural successor to SQL-based databases. Fortunately (for attackers), they are susceptible to many of the same injection-based attacks that befell their SQL-based predecessors.

As an example of a NoSQL injection vulnerability, let us revisit the Pixi application from earlier in the chapter. This uses MongoDB as the backing datastore, and the `/login` endpoint is susceptible to injection on the `pass` parameter, as shown in the following screenshot:

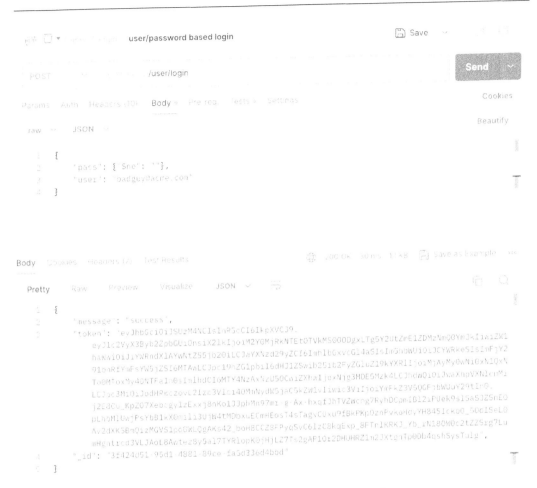

Figure 7.11 – NoSQL injection on the Pixi application

In this case, the endpoint has interpreted the NoSQL payload incorrectly, provided access to the user without a password, and returned an access token.

Command injection

A **command injection** is another nefarious form of injection attack that allows an attacker to trick the server into executing an operating system command at the behest of the attacker. A great example is the flaw in the home router (*#8: Home router* in *Chapter 4, Investigating Recent Breaches*), where an attacker could append a command to the specified IP address and cause the router to execute this command after the desired `ping` command. The attacker was able to inject a totally arbitrary command into the router and potentially take over the router entirely. This category of vulnerability is considered the most serious and frequently occurs with embedded systems. If you are working with an embedded system, always be on the lookout for operations (either in the web interfaces or

API) that will require the execution of some other processes (as in the `ping` example), as they may be vulnerable to injection attacks.

Path traversal

Path traversal (also known as directory traversal) is related to the command injection in so much as the attacker can manipulate the parameters passed to an underlying operating system command. In this case, it is the path (or directory) of the underlying resource (file, script, redirection, etc.) that can be manipulated by an attacker. This attack vector allows an attacker to deceive the server into accessing a resource (opening a file, for instance) in a location under the control of the attacker. This allows an attacker to potentially control the behavior of the server.

Typical attack patterns involved identifying resources that reference file paths and then modifying these to test a path traversal vulnerability. A typical example shows how the `. . /` traversal operator can be used to reference the root filesystem and then to traverse from there to a protected location:

```
filename=/var/www/images/../../../etc/passwd
```

Similar to command injection, this is a prized find for an attacker, as it facilitates a number of attack scenarios.

Server-side request forgery

The final injection we will cover is **server-side request forgery** (**SSRF**), which was included in the OWASP API Security Top 10 in 2023. The concept behind this attack is simple: an attacker can deceive the server into accessing a URL under the control of the attacker, thereby allowing the attacker to control the server. In this sense, it is very similar in impact to the path traversal attack described previously.

As an example, let us consider a server that accepts a request in the following format:

```
POST /product/stock HTTP/1.0
Content-Type: application/x-www-form-urlencoded
Content-Length: 118
basketApi=http://order.myshop.net/product/order/check_order
```

Now imagine an attacker can forge the request to the following form:

```
POST /product/stock HTTP/1.0
Content-Type: application/x-www-form-urlencoded
Content-Length: 118
basketApi=http://localhost/admin
```

This would allow the attacker to force the server to reference a file (potentially rogue) on the local filesystem.

Other API attacks

Finally, let us conclude this chapter with an overview of the catch-all API attacks not covered in one of the broader OWASP API Security Top 10 categories already covered.

API abuse

First, API abuse is considered an API attack that is not reliant on an actual vulnerability in the API itself but rather the abuse of the API, usually by using it in a way in which it was not intended. Typically, this attack vector includes excessive data exfiltration via APIs (think of scraping data from airline booking sites or estate agencies) or via discovering APIs' underlying websites and using them in ways that are not possible via the website.

Unrestricted access to sensitive business flows

This is a new vulnerability category in the OWASP API Security Top 10 in 2023 that, like API abuse, revolves around using an API (or a collection of APIs) to exploit the underlying business flow it enables. The canonical example is an online booking system whereby an attacker can exploit the logic of the business flow — a typical attack scenario allows an attacker to bulk book or reserve a ticketed item without payment, wait until demand escalates, and exploit the scarcity to their advantage. Typically, this technique is used by bots operating ticket scalping or resale sites.

This attack is difficult to prevent because it only becomes apparent after being executed at scale, by which stage it is already too late to prevent. For an attacker, this vector can be easy to discover simply by exploring the usual business flows and the underlying APIs and data flows and working out how to automate those to manipulate normal operations.

Business logic attacks

Finally, so-called **business logic attacks** cover other ways an API can be attacked. Typically, they focus on APIs that have some fragility in their underlying implementation and are reliant on their client utilizing the API in the prescribed manner. A tell-tale sign of a fragile implementation is an API that specifies a sequence of operations (do one thing before doing another), arbitrary restrictions (do not do this operation), or specific restrictions on input data (only a specific format of data is accepted). As an attacker, this is a red flag indicating that an API may be fragile and can easily be made to misoperate by ignoring instructions. For example, by ignoring a setup API call, it may expose some default behaviors that leak information, or by submitting data in a different format, the API may behave randomly, exposing further internal workings.

Summary

Congratulations on reaching the end of this chapter – we have covered a lot of in-depth information on the practicality of attacking an API. At this point, you should have an appreciation for the fundamentals of many of the different attack types, and hopefully, you're curious to try out some of these attacks yourself.

The attack techniques we covered align with the OWASP API Security Top 10 vulnerabilities, with a focus on authentication, authorization, and data-oriented attacks. By some measure, the most significant topic was the variety of ways in which authentication can be attacked and subverted in API systems — as a defender, this should be your first attack vector. Second only to authentication is the broad range of attacks that can be leveraged against authorization at both the object and function levels. We also learned key skills in attacking API data regarding both requests and responses. Finally, we covered a miscellany of attack types (injection attacks, business logic, and abuse attacks), which can be used when all else fails. For those really interested in learning more, the best way to start learning hands-on is in one of the many online laboratories.

This chapter also marks the conclusion of the second part of this book, namely *Attacking APIs*. From now on, we will switch our focus to defending APIs. In the next chapter, we look at defending against the very attacks we have covered in this chapter.

Further reading

These links provide further reading on password brute forcing and fuzzing:

- `https://portswigger.net/web-security/authentication/password-based`
- `https://github.com/0xZDH/burp-password-spray`
- `https://github.com/OJ/gobuster`
- `https://owasp.org/www-project-web-security-testing-guide/latest/4-Web_Application_Security_Testing/04-Authentication_Testing/09-Testing_for_Weak_Password_Change_or_Reset_Functionalities`

These links provide further reading on token attacks and manipulation:

- `https://www.elastic.co/blog/how-attackers-abuse-access-token-manipulation`
- `https://portswigger.net/web-security/jwt`
- `https://portswigger.net/web-security/jwt/algorithm-confusion`
- `https://danaepp.com/how-to-use-azure-to-crack-api-auth-tokens`

- `https://attack.mitre.org/techniques/T1134/`
- `https://www.howmanydayssinceajwtalgnonevuln.com/`

These links provide further reading on API keys and tokens:

- `https://nordicapis.com/why-api-keys-are-not-enough/`
- `https://nordicapis.com/the-difference-between-api-keys-and-api-tokens/`
- `https://blog.shiftleft.io/api-security-101-broken-user-authentication-1df2ef3420d8`
- `https://dzone.com/articles/security-best-practices-for-managing-api-access-to`
- `https://owasp.org/www-project-web-security-testing-guide/latest/4-Web_Application_Security_Testing/04-Authentication_Testing/09-Testing_for_Weak_Password_Change_or_Reset_Functionalities`

These links provide further reading on authorization attacks:

- `https://inonst.medium.com/a-deep-dive-on-the-most-critical-api-vulnerability-bola-1342224ec3f2`
- `https://www.bugcrowd.com/blog/how-to-find-idor-insecure-direct-object-reference-vulnerabilities-for-large-bounty-rewards/`

These links provide further reading on other attack types:

- `https://owasp.org/www-community/vulnerabilities/Business_logic_vulnerability`
- `https://portswigger.net/web-security/xxe`
- `https://portswigger.net/web-security/ssrf`
- `https://portswigger.net/web-security/sql-injection`
- `https://www.acunetix.com/vulnerabilities/web/rails-mass-assignment/`
- `https://cheatsheetseries.owasp.org/cheatsheets/Mass_Assignment_Cheat_Sheet.html`

Part 3: Defending APIs

In the final part, you will gain an in-depth understanding of how to defend your APIs by using tools and techniques across the development life cycle. The key sections in this part include how to shift left for API security, how to defend against common vulnerabilities, how to use best practices for securing your API frameworks and languages, and how to shield right by protecting APIs at runtime. We take a look at the future with a section on securing APIs in a microservices architecture and finally look at the key aspects of implementing an API security strategy.

This part has the following chapters:

- *Chapter 8, Shift Left for API Security*
- *Chapter 9, Defending against Common Vulnerabilities*
- *Chapter 10, Securing Your Frameworks and Languages*
- *Chapter 11, Shield Right for APIs with Runtime Protection*
- *Chapter 12, Securing Microservices*
- *Chapter 13, Implementing an API Security Strategy*

8

Shift-Left for API Security

Now that we have a deeper understanding of the methods and techniques used by API attackers, it is time to turn our focus to the core topic of defending APIs against attack.

In this opening chapter of *Part 3 – Defending APIs*, the focus will be on how API security can be improved by following a shift-left approach. The best way to avoid vulnerabilities in your APIs is to avoid introducing them in the first place. Although this sounds easier said than done, there is growing evidence from the software-development industry that addressing security concerns as early as possible in the lifecycle results in reduced risk and costs.

APIs are very well suited to a shift-left approach, since the appeal of a *design-first* approach (using the OpenAPI definition as the source of truth for the API) is fast becoming the de facto technique for API development. This chapter discusses various approaches to incorporating security during the design stage and then using the OpenAPI Specification to define security before the code is written so that it can be more easily enforced downstream through the software lifecycle.

In this chapter, we're going to cover the following main topics:

- Using the OpenAPI Specification
- Leveraging the positive security model
- Threat-modeling APIs
- Automating API security
- Thinking like an attacker

Technical requirements

For this chapter, you will need a development machine capable of the following:

- • Running Docker locally
- • Running VS Code with various marketplace extensions
- • Access to the internet and a GitHub account to access the examples

This chapter contains many code samples in various languages; these can either be run locally, which will require the installation of compilers, SDKs, and frameworks, or from within a Docker build container.

The example code and various breaking changes to the instructions can be found in the `Chapter 8` folder on the book's GitHub repository here: `https://github.com/PacktPublishing/Defending-APIs/tree/main/Chapter8`

Using the OpenAPI Specification

In *Chapter 1, What Is API Security?*, I described the various benefits of the *design-first* approach to API development: code generation, mock servers, data validation, documentation, and, of course, security (think of it as *security-as-code* for APIs).

We will now take a deeper look at the security constructs that can be leveraged within the OpenAPI definition of an API. There arc two main categories: data definitions and security specifications.

> **API-first versus design-first versus code-first**
>
> In this chapter, we will make heavy use of these three terms and, given that they have a fairly specific and implicit meaning in the world of API design, I would like to disambiguate them for readers who may not have this context:
>
> **API-first**: This is a business objective in which the product is built around APIs (which are invariably published) in the first instance and, from there, the user interface and so on is developed. This contrasts with the more familiar paradigm in which the user interface (web app or mobile app) is designed first, based on a monolith application or a set of internal APIs. The business exists to monetize its APIs, not its application. Twillio (`https://www.twilio.com/`) is often considered a canonical example of an API-first business.
>
> **Design-first**: This refers to a software-design methodology in which APIs are designed first using a specification language (most commonly Swagger or the OpenAPI Specification) and, subsequently, implemented as code. Software developers implement the API according to a design. Design-first offers numerous benefits, not least security by design, and is the preferred approach in this book.
>
> **Code-first**: This refers to a software-development process in which developers implement an API in code before a formal design is agreed upon. This may involve coding against a set of design-requirement documents or integrating with a working system. Code-first is considered something of a legacy approach, but it is perhaps the dominant approach, even in 2023.

Data

APIs are first and foremost conduits for data, allowing consumers to access data published by producers. It is vitally important that the producer fully understands their data sensitivity and does not inadvertently expose data that should be private, or otherwise leak data via the API. The key to being able to specify data accurately is the **data model** (schema), which defines the API data precisely by using data types, enumerations, dictionaries, arrays, and hashmaps.

The OpenAPI Specification defines the following primitive data types:

- `string` (including dates and files)
- `number`
- `integer`
- `boolean`
- `array`
- `object`

The OpenAPI Specification supports complex types, such as a dictionary, as shown here:

```
Components:
  schemas:
    Messages:              # <---- dictionary
      type: object
      additionalProperties:
        $ref: '#/components/schemas/Message'

    Message:
      type: object
      properties:
        code:
        type: integer
        text:
        type: string
```

This defines a dictionary called `Messages` with a string type as the key, and an object called `Message` as the value. The `additionalProperties` directive indicates that a dictionary is defined, and the use of the `$ref` directive indicates that the actual data type is defined elsewhere in the definition—think of this as a named type in a programming language. Rather than defining a `Message` instance multiple times, `$ref` can be used to refer to a common definition.

The `Message` type is defined as having two elements: a `code` value of the `integer` type, and a `text` value of the `string` type.

Different schemas may be combined by using one of the following directives: `oneOf`, `anyOf`, `allOf`, and `not`. This is best explained through the example here:

```
paths:
  /pets:
    patch:
      requestBody:
        content:
          application/json:
            schema:
              oneOf:
                - $ref: '#/components/schemas/Cat'
                - $ref: '#/components/schemas/Dog'
```

This specifies that the request's body may be *exactly* `oneOf` either `Cat` or `Dog`. No other response type is permitted.

The `Dog` type is defined in this snippet:

```
Components:
  schemas:
    Dog:
      type: object
      properties:
        bark:
          type: boolean
        breed:
          type: string
          enum: [Dingo, Husky, Retriever, Shepherd]
```

This example also illustrates the use of the `enum` directive, which constrains the `breed` to one of four possible options.

This is a brief description of how the OpenAPI Specification can be used to describe arbitrarily complex data structures; for full details on the OpenAPI Specification, refer to the current definition here: `https://spec.openapis.org/oas/v3.1.0`.

Let us now look at some definitions that have failed to fully specify their data types, leaving them vulnerable to attack or abuse. To this end, I will use the free 42Crunch VSCode Swagger editor extension (`https://marketplace.visualstudio.com/items?itemName=42Crunch.vscode-openapi`) to perform the audit of the definition, as shown here:

Figure 8.1 – Failure to constrain a string value

In this example, the API designer has specified the minimum and maximum lengths of the response string (which is definitely recommended); however, they have not constrained the string format. Let us assume a hypothetical example where a username uses the following format: USR-colin.domoney. By using regular expressions, we can create a definition of this username, as follows: ^USR-[a-z]+\. [a-z]+. This expression translates to: go to the start of the string, find the literal string USR-, find a character string (one or more lowercase letters comprising the *firstname*), find a . (full stop), and find a character string (one or more lowercase letters comprising the *lastname*). All very simple!

The specification can be amended to include this constraint by using the pattern directive, as shown here:

```
Chapter Nine example 1.json 9+, U  ×                              ⟳ ⥁ ◉ ◲ ▯ ⋯
paths ⟩ {} /users ⟩ {} get ⟩ {} responses ⟩ {} 200 ⟩ {} content ⟩ {} application/json ⟩ {} schema ⟩ {} properties ⟩ {} username
 21              "in": "query",                                                        ▯
 22              "name": "userID"
 23            }
 24          },
 25        "responses": {
 26          "200": {
 27            "description": "OK",                                                    ▯
 28            "content": {
 29              "application/json": {
 30                "schema": {                                                        ▯
 31                  "type": "object",
 32                  "properties": {
 33                    "username": {                                                  ▯
 34                      "type": "string",
 35                      "minLength": 6,
 36                      "maxLength": 255,                                            ▯
 37                      "pattern": "^USR-\\D*\\.\\D*"
 38                    },
 39                    "userID": {                                                     ─
 40                      "type": "integer",
 41                      "minimum": 1,
 42                      "maximum": 999,
 43                      "exclusiveMinimum": false
 44                    }
```

Figure 8.2 – Using the pattern directive

Other types of variable should be similarly constrained. Here is an example of an API that accepts an integer parameter but does not specify a minimum or maximum, as detected by the audit tool:

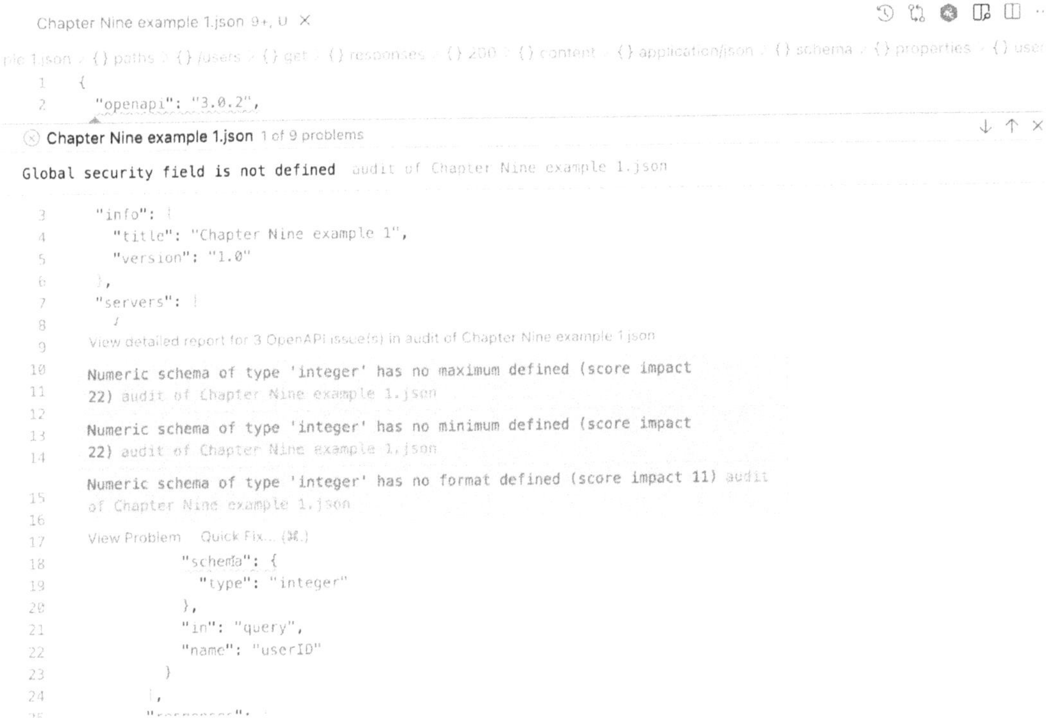

```
Chapter Nine example 1.json  9+, U  ×                                    ⟳  ⌕  ⚙  ⬒  ▯  ··
ple 1.json  ›  {} paths ›  {} /users ›  {} get ›  {} responses  ›  {} 200  ›  {} content  ›  {} application/json ›  {} schema ›  {} properties  ›  {} user
  1   {
  2     "openapi": "3.0.2",
                 ▲                                                                ↓  ↑  ×
⊘ Chapter Nine example 1.json  1 of 9 problems

Global security field is not defined  audit of Chapter Nine example 1.json

  3     "info": {
  4       "title": "Chapter Nine example 1",
  5       "version": "1.0"
  6     },
  7     "servers": {
  8       {
  9   View detailed report for 3 OpenAPI issue(s) in audit of Chapter Nine example 1.json

  10    Numeric schema of type 'integer' has no maximum defined (score impact
  11    22) audit of Chapter Nine example 1.json
  12
  13    Numeric schema of type 'integer' has no minimum defined (score impact
  14    22) audit of Chapter Nine example 1.json

  15    Numeric schema of type 'integer' has no format defined (score impact 11) audit
        of Chapter Nine example 1.json
  16
  17   View Problem   Quick Fix... (⌘.)
  18              "schema": {
  19                "type": "integer"
  20              },
  21              "in": "query",
  22              "name": "userID"
  23            }
  24          },
```

Figure 8.3 – Failure to constrain integer values

To establish the critical importance of constraining data, let us revisit the vulnerability affecting a dating application covered in *Chapter 4, Case Studies of Recent Breaches*. An attacker discovers that he can get high-precision geolocation information from the API and use this to triangulate a user's precise location. The root cause is the high-precision data (six decimal places) provided by the API, as follows:

```
{
  "user_id": 1234567890,
  "location": {
     «latitude»: 37.774904,
     «longitude»: 122.419422
  }
// ...etc...
}
```

By constraining the precision of the response with a pattern, this vulnerability could have been avoided by using an API firewall to block any response data not matching the contract or, even better, by testing the API fully before releasing it to production (this is discussed fully in *Chapter 12, Shield-Right for APIs with Runtime Protection*).

A reworking of the OpenAPI definition to include a `pattern` directive to constrain the geolocation values is shown here:

```
Chapter Nine example 2.json 8, U  ×

Chapter Nine >    Chapter Nine example 2.json > ...
  1    {
  2      "openapi": "3.0.2",
  3      "info": {
  4        "title": "Chapter Nine example 1",
  5        "version": "1.0"
  6      },
  7      "servers": [
  8        {
  9          "url": "https://api.server.test/v1"
 10        }
 11      ],
 12      "paths": {
 13        "/user_location": {
 14          "get": {
 15            "description": "",
 16            "parameters": [],
 17            "responses": {
 18              "200": {
 19                "description": "OK",
 20                "content": {
 21                  "application/json": {
 22                    "schema": {
 23                      "type": "object",
 24                      "properties": {
 25                        "latitude": {
 26                          "type": "string",
 27                          "pattern": "^\\d{1,3}\\.\\d{1}$"
 28                        },
 29                        "longitude": {
 30                          "type": "string",
 31                          "pattern": "^\\d{1,3}\\.\\d{1}$"
 32                        }
 33                      }
 34                    }
 35                  }
 36                }
 37              }
 38            }
 39          },
```

Figure 8.4 – Constraining excessive precision data

The regular expression used in this example is `^\d{1,3}\.\d{1}$`, which translates to: go to the start of the string, find at least one and at most three numbers, find a decimal point, find exactly one number, and find the end of the string. Our previous value of `37.774904` would fail to pass this test (due to the six digits of precision); however, a value of `37.7` would be accepted.

I have labored the point about fully specifying data types in the OpenAPI definition as this is a critical factor in API security. Even if your developers ignore the contract and write a backend that still returns excessive precision data, the API firewall can block this response if it violates the definition or detects the violation in the testing. The *definition is the contract*—anything not in the contract is invalid.

Auditing OpenAPI definitions in developer's IDEs

The example illustrations of OpenAPI definitions in Visual Studio Code are audited (or linted) using the 42Crunch audit plugin, which is available for free in the Microsoft marketplace. The plugins are available for free from here: `https://42crunch.com/resources-free-tools/`. OpenAPI definitions can be laborious to create from scratch, and readers should consider using a dedicated OpenAPI editor that supports the graphical editing and autogeneration of code elements. For Mac users, the Paw editor is recommended: `https://paw.cloud/`. StopLight is a good general-purpose tool: `https://stoplight.io/`.

Now that we understand the various directives relating to data elements within the OpenAPI definition, we will focus on the critical security directives.

Security

The OpenAPI Specification provides directives with which to specify various security mechanisms and their configuration within an API, specifically authentication and authorization.

The following schemes are supported:

- HTTP authentication schemes (including the basic and bearer types)
- API keys in headers, query strings, or cookies
- OAuth 2
- OpenID Connect Discovery

Security is described using the `securitySchemes` and `security` directives. The `securitySchemes` directive is used to specify all the security schemes your API supports (from the four possible types described previously), and then the `security` directive is used to apply specific schemes to the whole API or individual operations.

First, let us examine how the `securitySchemes` directive is used through the following example (in reality, it is unlikely that all five schemes would be used; this is purely for illustration purposes):

```
components:
  securitySchemes:
    BasicAuth:
      type: http
      scheme: basic
    BearerAuth:
      type: http
      scheme: bearer
    ApiKeyAuth:
      type: apiKey
      in: header
      name: X-API-Key
    OpenID:
      type: openIdConnect
      openIdConnectUrl: ...
    OAuth2:
      type: oauth2
      flows:
        authorizationCode:
          authorizationUrl: ...
          tokenUrl: ...
          scopes:
            read: Grants read access
            write: Grants write access
            admin: Grants access to admin operations
```

This API makes use of five variants for security, including `ApiKeyAuth`, where the location of the key is specified (in this case, `X-API-Key`). For OAuth2, two URLs are defined for the authorization and token, as well as the flow (in this case, `authorizationCode`).

The next step is to assign one of these security methods via the `security` directive to an endpoint to protect it, as follows:

```
paths:
  /billing_info:
    get:
      summary: Gets the account billing info
      security:
        - OAuth2: [admin]
      responses:
        '200':
```

```
          description: OK
        '401':
          description: Not authenticated
  /ping:
    get:
      security: []
      responses:
        '200':
          description: Server is up and running
        default:
          description: Something is wrong
```

The /billing_info endpoint is protected with OAuth2 within the admin scope, while the / ping endpoint is unprotected. This shows how easily these vital access-control directives can be added to an OpenAPI definition.

Similarly, it is a simple matter to audit an OpenAPI definition to detect whether there are any missing access controls. In the following example, the audit tool has detected that no global security schemes are defined within the definition, meaning that no access control is used in this API at all (note that the IDE plugin uses the Swagger/OpenAPI version 2 term securityDefinitions instead of the securitySchemes term used by OpenAPI version 3).

Figure 8.5 – Missing securitySchemes in OAS

In the following example, the audit tool has detected that the /users endpoint in the GET method is not protected by any form of access control.

Figure 8.6 – Missing security directive on endpoint

The API designer can easily remedy this by adding a security directive to the endpoint, as follows:

```
security:
  - OAuth2: [user]
```

The security directive can be applied within the local scope (per endpoint) or globally (for all endpoints, unless overridden). The best practice is to define a more restrictive policy globally and then relax this on a local scope, i.e., give a specific endpoint a null security definition. The benefit of this procedure arises in the event of the omission of a security directive, in which case, the more restrictive definition applies. This prevents endpoints from being left open and unsecured by accident. This is analogous to firewall rules—start with deny-all, then apply specific allow rules.

The final `security` directive that can be applied within an OpenAPI definition is transport security. In the following example, the server is specified as using the HTTP protocol, which is not considered best practice. The audit tool easily detects this and recommends the use of HTTPS instead.

```
Chapter Nine example 2.json 9, U  ×                                    ↺ ⟲ ⊙ ⊡ ⊟ ⋯
Chapter Nine     Chapter Nine example 2.json  { } servers
   1   {
   2      "openapi": "3.0.2",
   3      "info": {
   4         "title": "Chapter Nine example 1",
   5         "version": "1.0"
   6      },
   7      "servers": [
 Chapter Nine example 2.json  4 of 9 problems                               ↓ ↑ ×

API accepts HTTP requests in the clear  audit of Chapter Nine example 2.json

   8         {
   9            "url": "http://api.server.test/v1"
  10         }
  11      ],
  12      "paths": {
  13         "/user_location": {
```

Figure 8.7 – Detection of weak transport security

These simple examples show how easy it is to detect incomplete OpenAPI definitions at design time, and how easily these issues can be fixed before writing any code.

Generating client and server code

Now that we have a fully audited and validated OpenAPI definition, we can leverage the tremendous power of API code-generation tools to convert the definition into client-side code stubs, server-side code backends, or both. The use of this approach is referred to as the *design-first* approach—start with the API design and generate the code on that basis. The other approach is the *code-first* approach, in which developers produce an API first and then use introspection to reverse-engineer the API definition from the code.

While the design-first approach is the recommended method for new implementations, the code-first approach is probably still the most prevalent in the industry due to legacy APIs. It is possible to use a combination of the two approaches and then iterate between the two; for example, produce a version 1 of the API definition based on the **OpenAPI Standard (OAS)** definition, generate the code, test the API, make changes to the code, and reverse-engineer a version 2 of the API definition. This process is illustrated here:

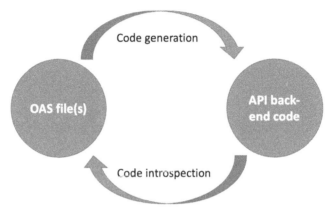

Figure 8.8 – Round-trip API development

The two most popular tools for the autogeneration of API code are as follows:

- Swagger Codegen (`https://swagger.io/tools/swagger-codegen/`)
- OpenAPI Generator (`https://openapi-generator.tech/`)

Both tools offer support for a wide range of popular languages and allow fine-grained control over the code generated to suit the developer. These tools will be covered in greater detail in *Chapter 11, Securing Your Frameworks and Languages*.

One of the biggest benefits of an OpenAPI definition is the ability within most API frameworks to automatically generate a portal within the application to allow consumers to experiment with the API before they develop their client application. This can lead to the easier adoption of APIs.

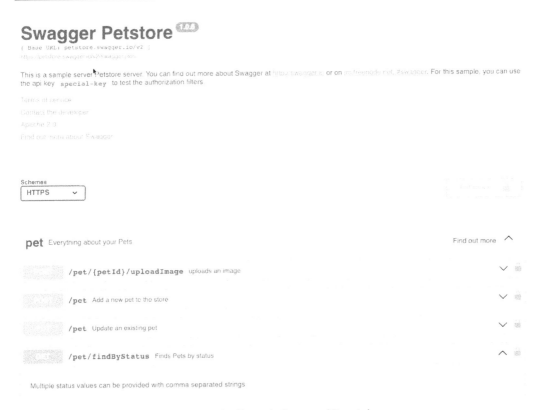

Figure 8.9 – Example Swagger UI portal

You can experiment with the popular Swagger `petstore` API here: `https://petstore.swagger.io/#/`.

> **Important note**
> This API definition is deliberately flawed for teaching purposes and should not be considered an example of good API design—it purely illustrates the Swagger UI.

Leveraging the positive security model

In the previous section, I made the following statement: *the definition is the contract*—anything not in the contract is invalid. This is the key benefit of the **positive security model**, which is paramount in the quest to produce secure APIs.

To understand the benefits of contract-based security, let us consider the alternative **negative security model**, or the so-called **blocklist** (or disallow list) approach. In this approach, a protection tool (such as a **Web Application Firewall (WAF)**) will have a list of malicious data and patterns, and block any requests containing such data.

To understand quite how fragile this approach is, let us look at a sample of the ruleset for ModSecurity (a popular WAF engine):

```
# Example Payloads Detected:
# -----------------------
# OR 1#
# DROP sampletable;--
# admin'--
# DROP/*comment*/sampletable
# DR/**/OP/*bypass blacklisting*/sampletable
# SELECT/*avoid-spaces*/password/**/FROM/**/Members
# SELECT /*!32302 1/0, */ 1 FROM tablename
# ' or 1=1#
# ' or 1=1-- -
# ' or 1=1/*
# 1='1' or-- -
# ' /*!50000or*/1='1
# ' /*!or*/1='1
# 0/**/union/*!50000select*/table_name`foo`/**/
```

While the few SQL Injection rules are valid, they are definitely not exhaustive. A skilled attacker could easily craft other payloads that would evade detection (for example, ' or 1=1;\x00). Ultimately, the strength of the WAF is limited by the skill of the rule administrator.

Although this model is conceptually simple, it is far from foolproof, and it suffers from three major drawbacks:

- Maintaining the blocklist is a laborious task. New items must be added continuously, and invalid items have to be removed. In addition, it is obviously impossible to know all possible malicious data, so the list is never going to be exhaustive.

- Since the ruleset can never be exhaustive, *false negatives* occur; in other words, valid attacks are not prevented as they should be.

- Finally, some rules can be overly restrictive, resulting in their triggering by valid data. These *false positives* can have a detrimental effect on the behavior of the application.

These three weaknesses have led to a widely held frustration with protection technology based on the **negative security model**.

The **positive security model** changes everything—we no longer attempt to block known malicious data and, instead, we only allow known valid data. This is a subtle distinction, but it is a critical one; instead of a **blocklist**, we now work based on an **allowlist**.

The immediate advantage of this approach is the precision of the protection—both false positives and false negatives are all but eliminated. There is one enormous caveat, though—the accuracy of this method is dependent on defining a precise contract for all data and operations.

Fortunately, for an API, we have just such a contract: our OpenAPI definition. If this has been well specified, it can be used as the source of truth for API protection based on a **positive security model**.

Conducting threat modeling of APIs

Another advantage of a shift-left approach to API development is that it allows security and development teams to participate in joint **threat modeling** activities. While a detailed description of threat modeling is beyond the scope of this book, the concept is a simple one (from the **Threat Modeling Manifesto**):

1. What are we working on?
2. What can go wrong?
3. What are we going to do about it?
4. Did we do a good enough job?

The value of threat modeling is best demonstrated with an example. Let us revisit the vulnerability affecting the website of a global shipping company covered in *Chapter 4*. The website developers made three basic errors in their design:

-   They relied on security by obscurity by returning a map image—the researchers were able to get an exact postcode from the maps by searching the street names

-   They did not rate-limit their tracking-query endpoint, which allowed the researchers to guess the parcel numbers using brute-force methods

-   They relied on client-side protections in the mobile application to protect the PII data exposed on the API

Two interesting observations strike me here: firstly, none of the issues are coding vulnerabilities, nor could they have been detected by scanning or auditing; and secondly, all three could have been prevented by simply asking a few questions at design time.

This is threat modeling in a nutshell. Is your assumption about a map not providing enough detail valid? How could an attacker abuse the tracking-query endpoint? Does the mobile application actually need PII information? All three of these issues could have been easily avoided.

There are other ancillary benefits to threat modeling, including greater cooperation between security and development teams.

Due to the rising interest in threat modeling as a practice, many excellent (free and paid) tools exist to automate the process. *Figure 9.10* shows a risk dashboard for an API with the popular **IriusRisk** (`https://www.iriusrisk.com/threat-modeling-platform`) tool.

Figure 8.10 – IriusRisk API threat model

In the preceding figure, the top panel shows a list of threats to a REST API, in this case, account or user enumeration through brute forcing. The lower panel shows three recommendations for protection or mitigation, the most obvious being rate-limiting.

The real power of using a tool is that threats and mitigations can be easily re-used between projects—by simply drag-and-dropping a **REST API** component, the user can pull in predefined threats and mitigations. This really is the power of collective wisdom in action.

Automating API security

When we consider all the possible ways in which an API can be vulnerable to attack, it may feel like a lost cause—with so many attack vectors, frameworks, access-control schemes, coding flaws, and so on, where do we even start?

The really good news is that a vast number of flaws are easily detectable using automated tools during the development lifecycle. A typical distribution of flaw occurrence (by count) against the difficulty of detection is shown here:

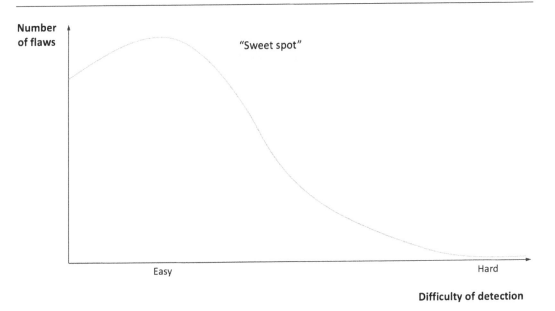

Figure 8.11 – Sweet spot for flaw detection

The *sweet spot* is where most of the low-hanging fruit exists—these are very basic flaws that can be detected easily. For example, all the OpenAPI definition flaws identified in the previous section can be detected in real time, with precision, at various stages in the lifecycle (IDEs, repositories, and CI/CD pipelines). Next, let us take a look at some quick wins from the integration API security checks into the development process.

CI/CD integration

For a successful CI/CD integration, a security tool should have the following characteristics:

- **Low latency of execution**: Security tools should not block the pipeline unnecessarily, and ideally for less than one minute
- **Low false positives**: Avoid breaking the build via flaws that are invalid or low severity
- **A good API**: Plugins should allow full automation via APIs; anything that requires a manual process (file download, etc.) is unsuitable for automation

Let us look at a few examples of how security testing can be automated in the CI/CD pipeline.

GitHub Actions

GitHub Actions allows CI/CD pipelines to be defined in code and executed automatically at key events in the code lifecycle. A typical example is automated audit scans on code commit or **pull requests** (**PRs**). The following screenshot shows how the 42Crunch audit tool can be included in GitHub:

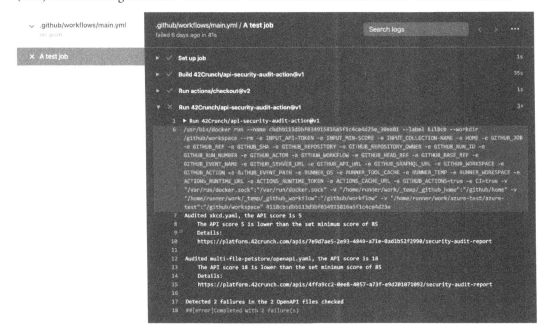

Figure 8.12 – Audit integration in GitHub Actions

Another example shows how **Semgrep** (covered in the next section) can be integrated into GitHub Actions:

```
code_analysis:
  runs-on:  ubuntu-latest
  name: Analyse code for security flaws
  steps:
    - uses: actions/checkout@v2
    - name: Code Security Analysis

    run: pip3 install semgrep && semgrep --config "p/ci"
    shell: bash
```

Jenkins

Jenkins is a stalwart of the software development industry and is still widely used. The following screenshot shows how the 42Crunch audit tool can be included in Jenkins:

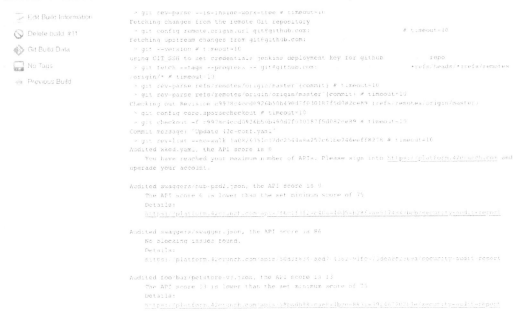

Figure 8.13 – Audit integration within Jenkins

Semgrep

The previous examples focused on the auditing of the OpenAPI definition, rather than on the resultant API backend code. It is possible to have a perfect OpenAPI definition and yet produce code that is vulnerable. Auditing both the definition and the code is essential for good security coverage.

Traditionally, the **static application security test** (**SAST**) has been the purview of well-funded security teams using high-end commercial solutions. Great advances have been made in the last few years toward lowering the barrier of entry to efficient and affordable static analysis. At the time of writing, the excellent **Semgrep** (`https://semgrep.dev/`) project was at the forefront of this charge.

Semgrep is a code-search tool (similar to the Unix **grep** utility), which has contextual awareness of code constructs such as functions, keywords, and classes. Semgrep offers exceptional search speeds and, by using its built-in rule syntax, users can construct complex queries to identify vulnerabilities in source code. There is a commercial version and a community edition, which comes with a large community-rule pack and a playground to create custom rules.

Let us look at how Semgrep can be used to detect basic API-coding flaws. A common bugbear of API developers is the correct use of **JSON Web Tokens** (**JWTs**), particularly their generation and validation. The Semgrep base ruleset has a built-in set of rules for assessing the correct usage of JWTs in a variety of popular languages.

The first scenario is the use of a JWT without verifying the signature, which presents the risk of tampering in transit. Without verifying the signature, the receiving party cannot be sure that the JWT has not been altered. In *Figure 9.14*, in the right panel, a vulnerable Java code snippet that attempts to use the JWT without first verifying it is shown. The left panel shows the rule that is highlighted, which looks for a call to `decode()` without a call to `verify()`. This rule can be incorporated anywhere within the lifecycle and takes a few seconds to run—this is how easily some very basic, yet severe, flaws can be avoided entirely.

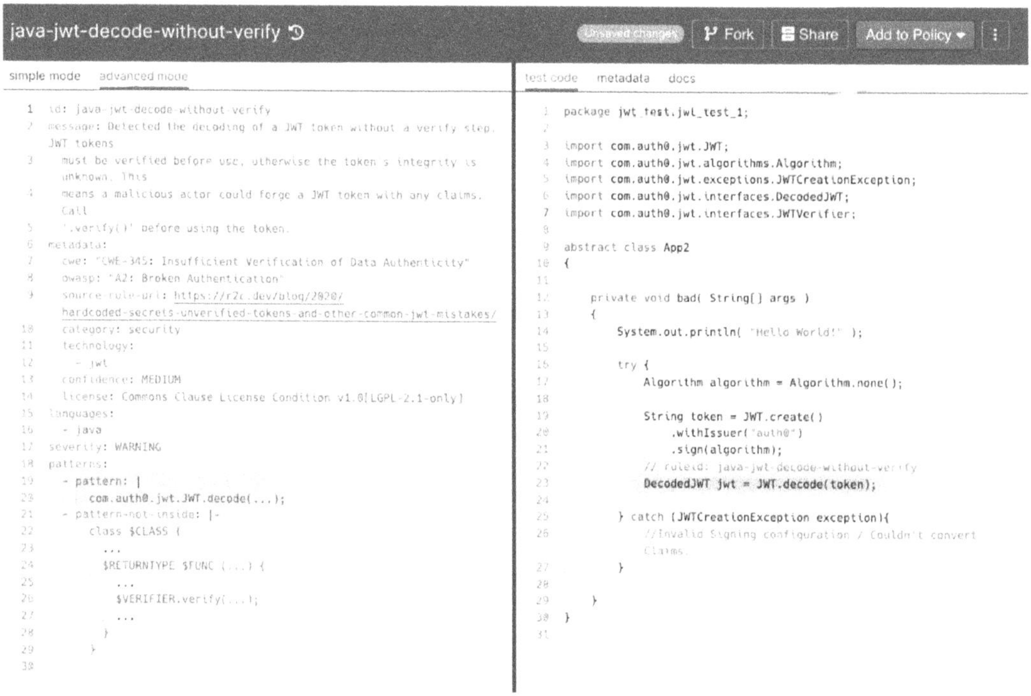

Figure 8.14 – Failure to validate the JWT signature

The second scenario is the failure to use a signing algorithm in the creation of the JWT, meaning the JWT cannot be verified—this is the `alg=none` vulnerability, which is so notorious that it even has its own website: `https://www.howmanydayssinceajwtalgnonevuln.com/`.

This is another flaw that is incredibly simple to detect with Semgrep, as follows:

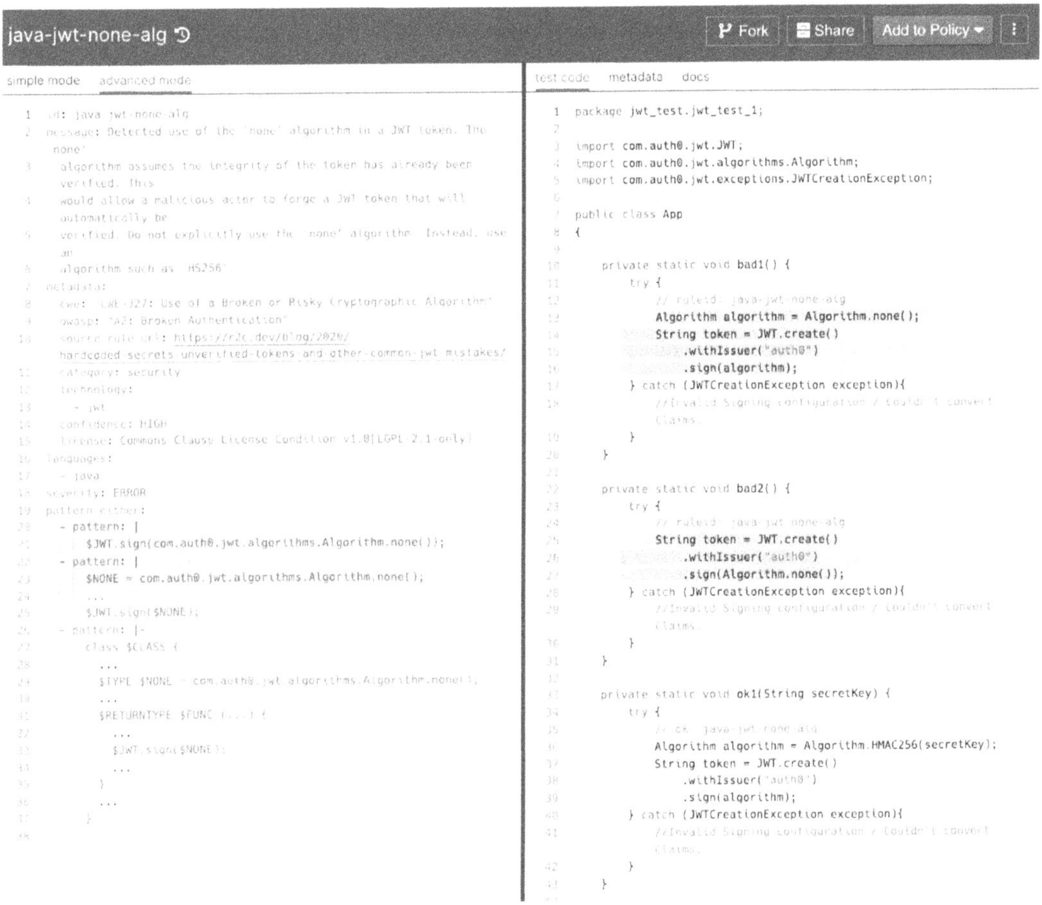

Figure 8.15 – Failure to use a cryptographic algorithm

This shows the immense power of an accurate, configurable, and fast tool—incorporate Semgrep into every phase of your lifecycle and eliminate this class of flaw forever. Similarly, if you have the luxury of a commercial SAST tool, make sure you are using this at every stage of the development lifecycle.

Thinking like an attacker

The most powerful piece of advice I could offer any API builder or defender is to think like an attacker. As discussed in the opening section, developers may not have a perspective on or insight into how an attacker can exploit an API. Unfortunately, as we have seen in *Chapter 4, Case Studies of Recent Breaches*, it is all too easy to become the victim of an API attack.

So, where does a developer start on their journey toward building more secure APIs?

- **Equip yourself with a toolbox and set of resources**: By far the most comprehensive list of API-security resources is the *Awesome API Security* GitHub repository: `https://github.com/arainho/awesome-api-security`.

- **Start with learning material**: There are great tutorials on YouTube, and an API-security course is even available for free online. OWASP as always is a fantastic reference.

- **Test your own APIs**: Use Postman and/or the reverse proxy tool to interact with your own APIs. Try to do things in an unexpected order, change some request parameters, and try unauthorized access.

- **Learn by doing**: There are some excellent, deliberately vulnerable API applications available. Install these, work through the tutorials, and exploit the APIs yourself.

- **Learn from others**: To me, the most useful lessons can be learned from the misfortune of others. Research and understand how API breaches occur and try to find the underlying issues. The APISecurity.io weekly newsletter contains in-depth coverage of recent breaches.

To get started, I would suggest running Postman against your own APIs and getting a thorough understanding of how the APIs work under the hood. Next, try and threat-model them to get an attacker's perspective.

Summary

We covered a lot in this chapter and, hopefully, you have a sense of optimism about the benefits of a shift-left approach toward API security. By embracing a design-first approach, development teams can incorporate security and data directives into their design at the very outset. By leveraging the OpenAPI definition as the single source of truth, developers can drive their development processes using API auditing and testing tools. The main benefit of this contract-based approach is the positive security model—no longer guessing what bad data looks like, we only allow good data. A predictor of success in the driving of a modern API development process is the tight integration of tools into the CI/CD pipelines, allowing the complete automation of the build, test, and deployment processes. Many of the most common vulnerabilities in APIs can be easily detected automatically and eliminated entirely—*this is the essence of the shift-left approach*. Finally, incorporating your security requirements (via threat modeling and security design) early in the process ensures that they receive due focus and attention during subsequent stages.

In the next chapter, we will move our focus to defending against the most common vulnerabilities affecting APIs.

Further reading

- For further information on threat modeling, the following are recommended:

  - `https://owasp.org/www-community/Threat_Modeling_Process`

  - `https://shostack.org/`

  - `https://www.iriusrisk.com/threat-modeling-platform`

  - `https://www.threatmodelingmanifesto.org/`

- For further information on ModSecurity setup, please refer to `https://www.digitalocean.com/community/tutorials/how-to-set-up-mod_security-with-apache-on-debian-ubuntu`

- The `Swagger.io` website provides a wealth of information on the OpenAPI Specification:

  - `https://swagger.io/blog/api-design/design-first-or-code-first-api-development/`

  - `https://swagger.io/docs/specification/data-models/`

- The Semgrep tool has detailed guides and a free version on its website: `https://semgrep.dev/`

- The Awesome API Security GitHub repository provides the single best resource for API security tools, methods, and technologies: `https://github.com/arainho/awesome-api-security`

9

Defending against Common Vulnerabilities

In this chapter, we take our first steps into learning how to defend APIs against common vulnerabilities in the design and development stage of the **Software Development Lifecycle** (**SDLC**). We have previously explored ways in which attackers can exploit weaknesses in API design and implementation and have reviewed past breaches for examples and fallout from insecure APIs; our focus now shifts to learning a defensive mindset to build secure APIs. This chapter will deal with each of the major classes of vulnerability types and, for each type, will provide best practices, common pitfalls, recommendations for tools and libraries, and code samples illustrating key defensive methods. If you are a developer, this is a key chapter in your learning journey, and by the end of this chapter, you will be well on your way to building secure APIs. For other readers, this chapter gives a solid understanding of key defensive techniques.

In *Chapter 7, Attacking APIs*, we learned how critical the twin pairing of authentication and authorization is for API security, and this core topic will form the focus of the first half of this chapter. The good news for API builders is that this topic is one of the easiest to improve by adopting a set of core practices, patterns, and supporting libraries. Data vulnerabilities can also be addressed using common defensive coding patterns, which you will master by the end of the chapter. Finally, we will take a broader look at defensive techniques to be used against a range of common implementation vulnerabilities and approaches to defend against business logic vulnerabilities at the implementation layer.

In a nutshell, this chapter is going to cover the following main topics:

- Authentication vulnerabilities
- Authorization vulnerabilities
- Data vulnerabilities
- Implementation vulnerabilities

- Protecting against unrestricted resource consumption

- Defending against API business-level attacks

We begin our exploration by examining how to defend against the twin challenges posed to API developers in building secure APIs, namely authentication and authorization.

Technical requirements

For this chapter, you will need a development machine capable of the following:

- Running VS Code with various marketplace extensions

- Accessing the internet and a GitHub account to access the examples

This chapter does not contain any specific code samples but provides code snippets and, in particular, JWT samples. No specific tools are required other than access to common online tools available via a standard web browser.

The example code and various breaking changes to the instructions can be found in the `Chapter 9` folder on the book's GitHub repository here: `https://github.com/PacktPublishing/Defending-APIs/tree/main/Chapter9`

Authentication vulnerabilities

Authentication attacks are the most frequently encountered attack vectors in APIs, and fortunately, they are also one of the easiest to defend by following core best practices in handling JWT security, implementing OAuth2 securely, and hardening your passwords, tokens, and your reset process.

Handling JWTs securely

In *Chapter 7, Token-Based Attacks,* we looked at various attacks against **JSON Web Tokens** (**JWTs**), since JWTs are ubiquitous within modern API implementations. Fortunately, nearly all of these attacks can be eliminated entirely by the secure handling of JWTs in the code that generates and consumes them.

The first recommendation is to *make sure you are using JWTs for their intended purpose* (a portable way of exchanging information about identity and permissions) and not attempting to use them where they are ill suited, for example, as a session cookie. Using this anti-pattern means a user cannot be logged out until the JWT expires, and JWTs used as session cookies can be stolen and re-used. In addition, JWTs are relatively bulky, leading to increased transmission times and storage. Additionally, cryptographic operations can be time-consuming if they need to be repeated on every request.

An important client-side consideration is that of *secure storage of JWTs* to prevent theft via **Cross-Site Scripting** (**XSS**) attacks. The safest storage location is browser memory or cache (known as session storage); however, this storage will be cleared on any refresh. Another option is to use a browser's local storage location; however, this is susceptible to XSS attacks. The current recommendation for secure token storage is to use the **HttpOnly** tag on the cookie location to prevent theft.

Another recommendation is to explicitly declare what your token will be used for in the `typ` field. For example, consider the JWT header in the following code snippet:

```
{
  "alg": "HS256",
  "typ": "JWT",
  "kid": "BTVBRYNjEyxc"}
```

With the type set to the generic default value of JWT, the client will have to make assumptions about the provenance of this token and potentially use it in an invalid context. By explicitly setting the type to a known media type, it can make it explicitly obvious how the JWT is intended to be handled. Using the previous example, a token can be declared as an identity token as shown here:

```
{
  "alg": "HS256",
  "typ": "IT+JWT",
  "kid": "BTVBRYNjEyxc"}
```

One of the most important properties of a JWT is the expiration timestamp to avoid tokens that are valid for much longer than intended. It is important that the JWT is generated with an appropriate expiration window and that the client validates the expiration before trusting the token – these two elements are fundamental to *handling expiration timestamps*. Suitable lifetimes for tokens will depend on the application; in the case of internal services, a few seconds may suffice, while for an external API requiring a human login, a much longer expiration (in the order of days) may be necessary. Unfortunately, there is no hard and fast rule here, and as usual with security design decisions, there is a trade-off between security and convenience. To determine suitable ranges for timeouts, consider performing threat modeling to consider worst-case scenarios of lost or stolen tokens and the relative difficulty in re-issuing a token.

Using algorithms securely

In *Chapter 7, Attacking APIs*, the *Token-based attacks* section showed various ways in which an attacker could inject keys and/or confuse the client about the signature algorithm in use. The simplest way to avoid this attack vector is to be explicit about the algorithm and key used and *ignore the values supplied in the JWT header*. Doing so removes the likelihood that an attacker can confuse the client about which algorithm to use or where to obtain the signing key. Although, at first glance, this may appear to remove some of the flexibility of the extensible, parameterized JWT architecture, this is offset since, in most cases, the JWT issuer and consumer are known to one another or at least well defined. Using hardcoded values is, in many cases, perfectly feasible.

Always, always ensure that JWTs are signed upon generation (using a suitable algorithm such as HS256) and that the signature is verified upon consumption.

Finally, when using symmetric keys, the secret key must be known to the client, and *due care should be exercised in distributing these keys to the client.* In particular, beware of many of the anti-patterns associated with secret distribution, such as storing secrets in plain text or committing them to source code repositories. Use a suitable key vault solution for your environment.

Using your libraries correctly

JWTs are actually very easy to generate and consume securely by following a few well-established patterns. The following code snippet shows a basic JWT token generation in Python using the standard jwt library:

```
def signJWT(user_id: str) -> Dict[str, str]:
    payload = {"user_id": user_id,
               "expires": time.time() + 600}
    token = jwt.encode(payload,
            JWT_SECRET,
            algorithm=JWT_ALGORITHM)
    return token_response(token)
```

The code shows how a single encode method performs the operation using the specified secret and algorithm (both values retrieved from secure storage), and an expiration window of 10 minutes.

The following is the corresponding client code:

```
def decodeJWT(token: str) -> dict:
    try:
        decoded_token = jwt.decode(token,
                JWT_SECRET,
                    algorithms=[JWT_ALGORITHM])
        return decoded_token if decoded_token["expires"]
                        >=    time.time()
                        else None
    except:
        return {}
```

The code shows how the decode method validates the token using the secret and algorithm (again provided from secure storage). In particular, as discussed earlier, the algorithm used is not taken from the JWT header but uses a hardcoded value. Although not explicit, the decode method also verifies the signature (always be sure to check the documentation regarding your particular library to ensure this is the case). Finally, the expiration window is checked; if the token is valid, it can be used; otherwise, a null value is returned.

When validating your JWTs, consider checking the headers for extraneous values that may have been inserted by an attacker or a third party. This is especially applicable if you are building a closed system where the client and server are using a priori values, this can be an easy indicator of compromise.

Testing your JWTs

Finally, some basic testing during development can greatly improve the secure handling of JWTs.

As part of the development process, unit tests should be rigorously performed on all functions involved in handling JWTs. A framework of validation tests should be used to check for a variety of corner cases, including `alg:none`, wrong key, wrong algorithm, missing fields, poor expiration window, and so on.

Similarly, using any static code analysis tools covered in *Chapter 8, Shift-Left for API Security*, can help discover basic implementation issues, such as failing to call the validate method or using unacceptable algorithms. These checks are simple to write, efficient to execute, and can catch various implementation issues.

Implementing OAuth2

In *Chapter 2, Understanding APIs*, the *OAuth 2.0* section explained the fundamentals of this protocol and its foundational importance in modern web applications and APIs. While the protocol is well established (currently in its second incarnation) and designed with security in mind, some anti-patterns can frequently weaken otherwise secure implementations. This section will cover recommendations for OAuth2 implementation for various application topologies.

The first recommendation is to use one of the preferred flows to suit your environment. If a human user is involved, this is the **Authorization Code Flow**. Every developer should have a solid understanding of this flow; the official OAuth2 playground is an excellent hands-on learning resource.

If the client is public (a web browser) or untrusted, then the extension to the Authorization Code Flow should be used; this is the **Authorization Code with Proof Key for Code Exchange** (**PKCE**). This flow extends the authorization code by adding a code challenge that prevents an attacker from intercepting the authorization code and exchanging it for an access token without the verification code.

Avoid using the **Implicit Flow** in new applications. Although this flow has no explicit vulnerabilities, it is no longer recommended as best practice by the IETF OAuth2 working group since it may be susceptible to token leakage and access token replays. The recommendation is to replace use with the PKCE.

For machine-to-machine communications, the **Client Credentials Grant Type** should be used (since the Authorization Code Flow requires the intervention of a human user to authorize the request). In the Client Credentials Grant Type, the *client secret* is exchanged for a (short-lived) token allowing access to the API. If the token is leaked, the impact can be minimized by setting a short expiration time (typically 24 hours or less) and limiting the scope of the claims. This makes this flow more flexible than using API keys that potentially have a broader scope and much longer lifetimes. The most important security requirement is to ensure that the *client secret* is kept secret and never transmitted or stored in publicly accessible storage such as a browser. It should be stored in a cryptographically secure manner in the client code or binary.

In *Chapter 4, Investigating Recent Breaches,* we saw an example of a smart scale leaking an OAuth2 refresh token in *#10: Smart scale.* A refresh token is highly sensitive since, with it, an attacker can create as many access tokens as he wishes until the refresh token is invalidated (this effectively eliminates the expiry window in the token since it can be regenerated when it expires using the refresh token). As such, the refresh token should be protected as far as possible. On the client side, ensure that the refresh token is stored securely, and for frontend web clients without secure storage, ensure they are short-lived to mitigate damage from leakage. On the server side, ensure that refresh tokens are invalidated if a user explicitly logs out of a session or changes their password. On the server, be sure to monitor for attempts to use invalidated tokens since this may be an indication of an attack in progress.

Since one of the design goals of OAuth2 is that of delegated authorization, ensure that the minimum level of access is granted to your tokens. Avoid issuing tokens with excessive scope and use the audience field to minimize its scope further. Also, use the expiration period to minimize the impact of token leakage.

One of the most common vectors for token leakage is via XSS vulnerabilities on the frontend. Ensure that every effort is made to eliminate XSS vulnerabilities and that **Cross-Origin Resource Sharing (CORS)** has been set up correctly on your API to minimize the likelihood of token leakage.

Finally, there are some excellent references in the *Further reading* section. In particular, I recommend the *Pragmatic Web Security* best practices guide and the two IETF articles on best practices and threat modeling.

Password and token hardening

The recommendations for improving passwords and token security are twofold: increase the complexity (permutations of possible characters) and length (number of possible characters) of the password or token. This is one of the most deterministic improvements that can be made toward improving security, and there should be no conceivable reason not to use long, complex passwords and tokens.

If a human user is involved in the authentication process, they may resist the prospect of using long, complex passwords. In this case, a secure password storage solution, such as 1Password, LastPass, or similar, should be used. Certainly, from a security perspective, the prospect of moderate user inconvenience is preferable to using weak passwords, which – as we have seen – can be relatively easily cracked.

The time taken to crack a password or token increases exponentially as a function of complexity and length, as shown in the following table (taken from data in 2023):

Number of characters	Complexity	Time taken
4	Mixed case letters	Instantly
8	Mixed case letters	28 seconds
12	Mixed case letters	6 years
14	Mixed case letters	17,000 years
16	Numbers and mixed case letters	779 million years
18	Numbers and mixed case letters	2 trillion years

Table 9.1 – Time taken to crack passwords or tokens

In 2023, a password or token with a length of 16 or more characters and with complex characters should be resistant within practical reason from the attacks covered in *Chapter 7, Attacking APIs*.

The other recommendation for generating passwords and tokens is to use a pseudo-random number generator with a high degree of entropy to avoid producing passwords or tokens with a predictable character sequence. Different programming languages will have different standard methods available, so be sure to consult the reference manuals to find one that is recommended to be *cryptographically secure*. In the case of Java, it is preferable to use the `SecureRandom()` class over the `Random()` class.

Securing the reset process

In *Chapter 4, Investigating Recent Breaches,* we saw an example of a smart scale with a broken password reset process in *#10: Smart scale*. In this example, the reset PIN was only four digits long and in a very narrow range of characters, meaning it was easily guessable by the researchers. Additionally, a reset token was never invalidated, which meant it could be re-used to attempt further password reset.

Due to the critical nature of a password reset process, great care should be taken when designing and implementing the process. Certainly, threat modeling should be used to understand possible abuse and corner cases and ensure mitigations are in place against these.

The following best practices are recommended for password reset processes:

- Ensure that the time window for the start-to-finish of the process is bounded and that on timeout, the process is reset entirely, including invalidating any tokens or PINs in use
- Use a trusted side channel (typically email) to communicate the reset sequence
- Ensure that responses are uniform in content and timing, whether or not the account exists, to prevent enumeration attacks

- Log and track suspicious reset activity for potential incident detection and response

- Ensure that reset PINs or tokens are sufficiently random and cannot be easily guessed

- Implement progressive rate-limiting on the reset request endpoint to prevent an attacker from brute-forcing the reset process

- Use a **step-up process** (such as enforcing additional security factors such as a security question) if a user performs multiple reset attempts or enters incorrect information

- Always confirm the user at the start and end of the process to indicate a reset is in process, giving them the option to abort it if they did not initiate it

Once implemented, the password reset sequence should be tested thoroughly for both normal and abuse cases, ideally by a skilled **red team** or external penetration tester skilled at exploiting such systems.

Handling authentication in code

Finally, to wrap up this section on authentication, we will look at some examples of how to add authentication to an API endpoint. Firstly, we will look at an example in **Python 3** with the **FastAPI** framework, and secondly, in **Node.js** with the **Express** framework.

For FastAPI, this is achieved using dependency injection to inject an authentication handler into the API endpoint handler as shown in the following code block:

```
@app.post("/posts", dependencies=[Depends(JWTBearer())],
tags=["posts"])
async def add_post(post: PostSchema) -> dict:
    post.id = len(posts) + 1
    posts.append(post.dict())
    return {
      "data": "post added."
    }
```

The authentication validation is performed in the JWTBearer() class as follows:

```
class JWTBearer(HTTPBearer):
    async def __call__(self, request: Request):
    credentials: HTTPAuthorizationCredentials =
        await super(JWTBearer, self).__call__(request)
    if credentials:
        if not credentials.scheme == "Bearer":
            raise HTTPException(status_code=403, detail="Invalid
authentication scheme.")
        if not self.verify_jwt(credentials.credentials):
            raise HTTPException(status_code=403, detail="Invalid token
or expired token.")
```

```
        return credentials.credentials
    else:
        raise HTTPException(status_code=403, detail="Invalid
authorization code.")
```

The code performs the following checks:

1. Checks whether the credential is supplied at all, and if not, raises an exception with the message `"Invalid authorization code."`.

2. Checks whether the credentials supplied are of the 'Bearer' type, and if not, raises an exception with the message `"Invalid authentication scheme."`.

3. Finally, it checks whether the JWT token is valid, and if not, raises an exception with the message `"Invalid token or expired token."`.

4. If no exceptions have been raised, the function returns the credentials to the API handler, where execution continues as normal, having been authenticated.

This example shows how a relatively small amount of code is required to handle authentication correctly and securely – the key is to ensure that the authentication handler is injected into every API endpoint.

As an illustration, let us look at how the same task is accomplished using Node.js with the Express framework. Firstly, the following code shows how the `isAuth` method is added to the call list in the API endpoint handler:

```
app.get("/secrets", isAuth , (req,res) => {
    const secrets = [{
        id: 1,
        name: "Secret 1"
        }];
    res.json(secrets);
});
```

The `isAuth` implementation is shown in the following code block:

```
function isAuth(req, res, next) {
    const auth = req.headers.authorization;
    if (auth === 'password') {
        next();
    } else {
        res.status(401);
        res.send('Access forbidden');
    }
}
```

In this example, the authentication handler performs a very basic check against a password. In the event of a mismatch, a **401 status** code is returned together with a message of **Access forbidden**.

> **About the code sample**
>
> This chapter (and the upcoming *Chapters 10*, *11*, and *12*) will feature code samples to illustrate key learning points. Due to the variety of programming languages and frameworks, I cannot provide code samples to suit every possible audience. I have opted to use **Python 3** as a reference language (since its syntax is closest to natural language and probably most widely readable) and the **FastAPI** framework (a modern API framework for Python). I apologize in advance to readers unfamiliar with either of those and hope this does not detract from your learning experience.
>
> As the book evolves, I expect to provide other code samples in other popular languages and frameworks; these will be published in the books accompanying GitHub repository:
>
> ```
> https://github.com/PacktPublishing/Defending-APIs
> ```

Authorization vulnerabilities

Now that we have covered how to secure your API authentication, we focus on its counterpart – authorization. We will cover patterns for protecting against object-level and function-level vulnerabilities and how to apply various authorization middleware to improve overall authorization robustness and extensibility.

Object-level vulnerabilities

In *Chapter 3, Understanding Common API Vulnerabilities*, we covered the root causes of broken object-level vulnerabilities in the *API1:2019 – Broken object-level authorization* section. As a reminder, this vulnerability originates when an API grants access to an object (typically data) not owned by the calling user or client.

Despite its prevalence and reputation as the most serious of API vulnerabilities, broken object-level authorization is paradoxically one of the easiest vulnerabilities to address as a defender. The rule is simple – always explicitly validate the access to an object. Do not trust the rights of an existing session identifier, an identifier passed as a parameter, or a JWT, for instance.

This concept is perfectly illustrated by the code sample given in *Chapter 3* , *Understanding Common API*, repeated here:

```
Class UserController < ApplicationController
  def show
    if Autorization.user_has_access(current_user,
      params[:id])
        @this_user = User.find(params[:id])
```

```
        render json: @this_user
    end
  end
```

The highlighted one line of code is all that is required to eliminate this API handler's broken object-level vulnerabilities. Here, the code explicitly validates whether `current_user` (obtained via the authentication handler in the API middleware) has access to the parameter supplied in the `id` value.

An insecure implementation would have likely omitted this step, assuming the client-provided `id` value passed to this call was established in a prior call. For a great example of where an API failed to correctly validate a supplied object identifier, refer to *Chapter 4, Investigating Recent Breaches*, specifically the *#4: Cryptocurrency portal* section. In this example, there were two calls: first, to set up the transaction and then to execute the transaction. Unfortunately, in the second call, the API trusted the supplied account ID, allowing an attacker to switch the account ID to an account lacking funds. If implemented correctly, the API would have explicitly revalidated the source account and rejected it due to insufficient funds.

The other key defensive activity in securing against broken object-level authorization is comprehensive and specific testing for this type of vulnerability. Unfortunately, most current test frameworks are not designed to perform the multistage A-B testing required to identify a broken object-level vulnerability. I would recommend creating bespoke tests (using a Python or Node.js unit testing framework) to perform specific testing for this vulnerability type. It can usually be quite easily identified, and the investment pays off given the severity.

Function-level vulnerabilities

Similarly, broken function-level vulnerabilities can be relatively easily eliminated by applying secure coding patterns. Exactly the same guidance applies to object-level vulnerabilities – always explicitly validate the access rights of the user or client. In the case of function-level vulnerabilities, the check should be made against whether the caller has the right to access the function (as opposed to the object). A typical example is a client accessing a higher privilege endpoint such as a `/admin` endpoint allowing protected operations.

The following Python/FastAPI code shows how an authorization check could be made to validate a user's access to a privileged endpoint:

```
oauth2_scheme = Oauth2PasswordBearer(tokenUrl="token")
    ...
async def get_current_user(token: str = Depends(oauth2_scheme)):
    ...
@app.g"t("/ad"in")
async def do_admin(current_user: User = Depends(get_current_active_
user)):
    if not AuthZ.user_has_admin_access(current_user):
```

```
        raise HTTPException(status_code=401, deta"l="User does not
have admin privileg"s.")
    else:
        # Perform admin operations here
        pass
```

The highlighted code checks whether `current_user` has access to administrator privileges using the `user_has_admin_access()` method. If not, it raises an exception with the message " `User does not have admin privileges.`". If the user has the requisite privileges, then execution proceeds.

This shows how conceptually simple it is to address broken function-level vulnerabilities in code. Additional best practices include the following:

- Implement the access control logic at the top of the API handler method (as shown in this example) to protect the entire method.

- Make sure that any default behaviors should deny access rather than granting them. This ensures that coding errors or unhandled exceptions do not inadvertently grant access.

- Derive your access control method from an abstract base class that implements standard checks based on the user's group/roles, perhaps derived from central access control systems such as LDAP or Active Directory.

Similar to the case with object-level vulnerabilities, a well-written testing framework can easily identify function-level vulnerabilities. As a further note, be especially cautious of endpoints that accept both an object identifier and expose a high-privilege method since this may be vulnerable to both object- and function-level vulnerabilities.

Using authorization middleware

The previous discussion on object-level and function-level vulnerabilities has claimed that they are relatively easily addressed in code, as shown by the relatively straightforward code samples. However, this only tells part of the story. The biggest challenge with authorization is how to manage it at any scale in an application. Typical anti-patterns include solutions that are very fragile due to hard coding of the authorization logic (for example, the code samples above on object and function level vulnerabilites) and totally lacking extensibility (for example, how to extend the policies or adjust the permissions without having to rewrite the code).

This is a classic software engineering problem of excessively tightly coupled systems and is a major cause as to why authorization in practice is a much harder problem to solve than it would initially seem. Fortunately, there are a number of mature and well-designed **authorization frameworks** available to system designers that abstract the logic and policy frameworks from the end user.

Figure 9.1 shows the typical abstract architecture of an authorization framework (also called policy engines):

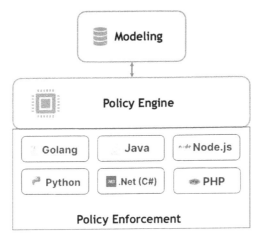

Figure 9.1 – High-level architecture of an authorization framework

The three key elements are as follows:

- **Modeling**: At the highest level the modeling component allows system designers to model their domain in terms of users, groups, roles, and permissions.

- **Policy engine**: The policy engine is where the logic of the authorization engine is implemented. For a given access request, the engine will determine whether access will be granted, based on the user, the resource, and their permission. Engines can support various models such as RBAC, ABAC, and a simple **Access Control List (ACL)**.

- **Policy enforcement**: Finally, policy enforcement is a client library integrated into the target platform to handle policy decisions. This code resembles the code samples we saw in the preceding sections, and instead of making decisions inline in code, the enforcement code will delegate the decision to the policy engine.

There are many excellent authorization frameworks available; here is a selection of the most commonly encountered choices:

- **Open Policy Agent (OPA)** (`https://www.openpolicyagent.org/`): OPA is one of the best known and is a **Cloud Native Computing Foundation (CNCF)** project. Strictly speaking, this is more of a policy engine and lacks direct support of the enforcement layer.

- **Oso** (`https://www.osohq.com/`): Oso is an increasingly popular choice and is billed as the *batteries-included* solution supporting most authorization patterns. It comes with a wide range of enforcement libraries and uses its own modeling language (called Polar) for policy definition. Oso also offers excellent educational content on how to implement authorization in practice.

- **Casbin** (https://casbin.org/): Casbin is a well-designed and lightweight authorization framework supporting most authorization patterns and many common enforcement libraries. Casbin uses a simple yet powerful modeling engine using a definable model and policy, both of which are interpreted at runtime.

Let us conclude this section with a deeper look into Casbin to understand how it works in a real-world example. Firstly, policies can be written and tested by a web-based designer as shown in *Figure 9.2*:

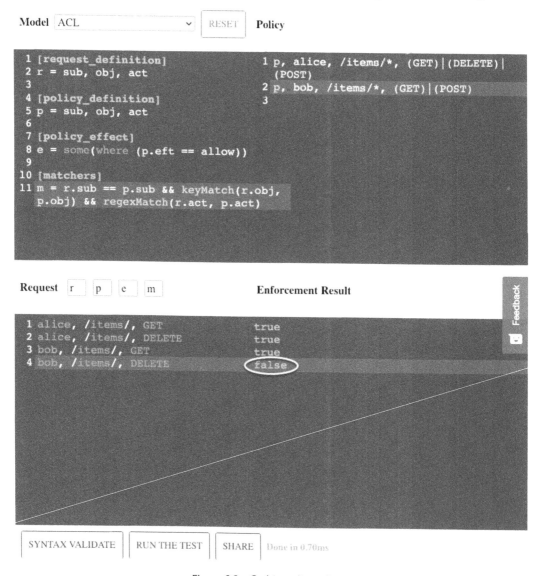

Figure 9.2 – Casbin policy editor

The three elements are as follows:

- In the top-left window is the **model definition**; in this case, a simple ACL model is shown
- In the top-right window is the **policy definition**; in this case, we have two users with some basic HTTP access permissions
- In the bottom window is the **test result viewer**, where a designer can provide various test scenarios and evaluate the outcome based on the model and the policy

To understand how the model and the policy work, we need to understand some key terms:

- r: The request definition; in this case, comprising a subject, an object, and a desired action
- p: This is the policy, which again comprises a subject, an object, and a desired action
- e: The effect is a logical expression of how the policy is evaluated
- m: This is the combinatorial logic that determines how a request is matched to a policy

To see how these are used to produce a working engine, consider the policy shown (top-right window) with the test data (four items in the bottom window). Looking at the policy for alice, she has access to the GET, DELETE, and POST methods on the /items/* endpoint, so she is granted access for her two test cases.

Similarly, bob has access to the GET method and is granted access. However, he does not have access to the DELETE method and is denied access. This is a simple example but shows the core concepts. The policy models and definitions are very extensible and can implement many real-world scenarios.

To see the power of Casbin in action, let us conclude by looking at the code implementation for the policy (in *Figure 9.2*) shown in Python 3/FastAPI.

Firstly, here is the code to invoke the policy engine and make a decision based on the requestor, the target URL, and the operation:

```
async def get_current_user_authorization(req: Request, curr_user: User
= Depends(get_current_active_user)):
        e = casbin.Enforcer("model.conf", "policy.csv")
        sub = curr_user.username
        obj = req.url.path
        act = req.method
        if not(e.enforce(sub, obj, act)):
            raise HTTPException(
                status_code=status.HTTP_401_UNAUTHORIZED,
                detail="Method not authorized for this user")
        return curr_user
```

The first point to notice is that the Casbin engine is instantiated at runtime from static data (a model configuration and a policy CSV file). This means that the policies can easily be modified by editing these two text files or pulled from a central location. Most importantly, the code does not have to be rebuilt, and the policy definition is now decoupled from the application code.

Secondly, notice how the subject, object, and actions are assigned to various API request objects; this again facilitates low coupling between policy and code.

Finally, let us look at the API endpoint for the DELETE method:

```
@app.delete("/items/{item_id}", status_code=status.HTTP_204_NO_
CONTENT)
async def delete_item(item_id: int, req: Request, curr_user: User =
Depends(get_current_user_authorization)):
    items_dao.delete_item(item_id)
    return Response(status_code=status.HTTP_204_NO_CONTENT)
```

The most notable aspect of this implementation is how concise and uncluttered the code is – all that is needed to implement the authorization controller logic is a single piece of *syntactical sugar* to add a dependency to the get_current_user_authorization method, which then handles all the authorization logic. In particular, note that there is no decision logic based on hardcoded names or variables in the code. The API developer never has even to consider what checks they to make regarding access; instead, they delegate this to the authorization layer handled by the Casbin middleware. All that they have to do is to remember to inject the appropriate method to invoke the middleware. Finally, it is a simple task to be able to run static analysis on the API code to check that this middleware was correctly invoked.

In conclusion, this code sample shows the power of using authorization middleware:

- Rules and policies can be decoupled from the API code itself
- Rules and policies can be dynamically adjusted and disseminated from a central server
- Standard access methods make the API code easier to write since the complexity is hidden from the API developer
- Standard access methods facilitate easy code reviews and automated testing to ensure the authorization code is correctly invoked

Using one of the many authorization controller implementations available, combined with an appropriate access model and policies, can ensure that many of the most common object-level and function-level authorization vulnerabilities can be addressed.

Data vulnerabilities

Data vulnerabilities are one of the most significant weaknesses impacting API security, with nearly all breaches involving data leakage to some extent. For API defenders, the good news is that it is a vulnerability class that can be defended using some core principles and techniques.

Let us start our journey by understanding how data propagates through an API from the request, via the API layer, then the database layer, where it will be persisted to a database storage layer. A response follows the reverse flow: data is accessed from the database via the database layer, processed by the API layer, and returned to the user or client in the response.

This is summarized in the following simplified architecture diagram:

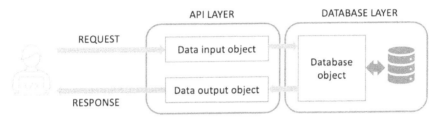

Figure 9.3 – Typical API data controller model

There are three data processing layers, each with its own data object:

- **Data input object**: This is the native input format data received in the API request, most commonly JSON, XML, or text

- **Database object**: This is the data format stored in the database, and may include database fields such as primary and foreign keys and indexes

- **Data output object**: This is the native output format data transmitted in the API response, most commonly JSON, XML, or text

Here is the vital learning point: API data vulnerabilities are usually caused when a developer maps *directly* from one data object to another without considering the sensitivity of the data. For example, if a database has a value for a hashed password, this should not ever be returned in an API response. Similarly, if an API request attempts to set a value called is_admin, the database layer should probably not write that to the database automatically, and a log event should be registered to indicate suspicious activity.

This is a really important concept to grasp. The API acts as a **translation layer** between the user and the database and should apply a set of translation rules in both directions rather than acting as a perfect conduit.

Let us examine the two data vulnerabilities in detail to understand how to implement this concept to protect data.

Excessive data exposure

Excessive data exposure occurs when an API returns more data than intended or is desirable. This is the most commonly occurring API vulnerability, and in *Chapter 4, Investigating Recent Breaches*, almost every vulnerability described featured this vulnerability either as the primary or secondary weakness.

There are two main areas to address excessive data exposure: at the code level and at the data classification level.

Coding securely to prevent excessive data exposure

For a developer, one of the easiest ways to convert a binary data object retrieved from a database into a format transmissible in an API response is to use one of the coalescing operators on the data object, such as the `to_json()` or `to_string()` helper methods. While these are extremely helpful in viewing data for debugging or converting to other intermediary formats, they can have unintended consequences if used without caution and should generally be avoided when constructing API request and response data. For example, a database may store highly sensitive user information about a user, which could be divulged if a `to_json()` method was used carelessly. Rather, be explicit about what fields will be mapped over, even if this means writing more code.

The other culprit is the **Object-Relational Mapper (ORM)**, a layer between the database data object and the raw database. Think of this as a helper layer hiding the database's implementation detail from the developer. The developer can write to a data object, and the ORM will format, index, and commit this to the database. Unfortunately, the ORM is unaware of the sensitivity of data fields and will return all fields, which can easily lead to leakage in the same manner as the `to_json()` method.

Let us return to *Figure 9.3* and examine how to implement the three different data objects securely. In this example, I will demonstrate how to use **Data Transfer Objects (DTOs)** in FastAPI to prevent excessive information exposure and mass assignment. For our API method, we will need a data object for the input data, including a password, a data object for the database including a hash, and an output object without any password information.

Here is some code to show the basic data models:

```
class UserBase(BaseModel):
    username: str
    email: EmailStr
    full_name: str | None = None

class UserIn(UserBase):
    password: str

class UserOut(UserBase):
    pass
```

```
class UserInDB(UserBase):
    hashed_password: str
```

UserBase contains the common data types present in all objects; the three other types (UserIn, UserOut, and UserInDB) extend this base class to add their required fields.

Using Python dictionary operators, it is simple to construct a derived object (in this case a UserInDB) from the base object:

```
user_in_db = UserInDB(**user_in.dict(),   hashed_
password=md5sum(password))
```

This is an excellent compromise between development efficiency and security: the developer is not forced to copy individual fields by hand (the dict() operator allows efficient object initialization), and additional fields must be explicitly added and initialized.

Classifying data to avoid ambiguity

In *Chapter 4, Investigating Recent Breaches*, the *#3: Microbrewery application* section showed an example where an API was returning more data than was necessary for its intended purpose. In the example we examined, this additional data contained PII data not required; however, it is quite possible that other client applications with a legitimate requirement for this data used this same API. This indicates another common anti-pattern: a single API being overly verbose, serving multiple clients. Rather than attempting to cover all options with a single API, the API teams should provide multiple APIs with different datasets. Most importantly, access to these datasets should be tightly regulated in accordance with their data classification.

Data types should be classified according to their sensitivity and intended audience. When a client application requests access to this API, a review is conducted into the legitimate business case for such access rather than having it granted by default. This is analogous to the principle of least privilege – unless you need the data, you do not get access to it.

Once the API data classifications have been determined by the business, risk, and compliance teams, then the API test fixtures should be enhanced to ensure that they are monitoring for data types outside of the scope of the API and flagging this as a possible data leakage failure. It is relatively easy to write tests that will identify PII data such as phone numbers, social security numbers, or credit card numbers in API responses.

In *Chapter 8, Shift-Left for API Security,* the *Using the OpenAPI Specification* section described how API data could be described using a schema within the API definition. This schema should be updated to specify the allowable data types for all response records, and test tools should test according to this contract. This topic will be explored fully in *Chapter 11, Shield Right for APIs with Runtime Protection.*

Mass assignment

Mass assignment is the slightly less common counterpart of excessive data exposure and occurs when an API accepts more data from the API than it should. Typically, an attacker will try and guess the names of sensitive field values and add these to the request payload, hoping that the API implementation will accept these extra values and include them within the database object. Using this technique, attackers can modify fields including access roles, passwords, or other reset information.

Like the case of excessive data exposure, the most common underlying cause of mass assignment is the use of ORMs, which bind input fields to the corresponding database field name. It is easy for an attacker to build up a word list of commonly used input fields to use in attacks based on common frameworks and ORMs. Common values include `is_admin`, `admin`, `privileged`, and `debug`. The same recommendation applies here as in the case of excessive information exposure – be specific about the values that can be written to the database object and avoid any implicit assignments.

The code sample repeated here from earlier shows this defensive coding practice very clearly:

```
user_in_db = UserInDB(**user_in.dict(),   hashed_
password=md5sum(password))
```

This code is entirely resistant to mass assignment attacks. Firstly, the `UserIn` class is designed to contain only non-sensitive values, and the only field written to the database, in addition, is the calculated password hash. The decision to add fields is explicit and deliberate, avoiding any possibility of implicit assignment.

Using input filtering (in a WAF) to strip any well-known fields in request input is also possible. This block-list approach is somewhat crude and may lead to false negatives (stripping actual valid data), but it is very effective against commonly used field types.

In *Chapter 8, Shift-Left for API Security*, the *Using the OpenAPI Specification* section described how API data could be described using a schema within the API definition. This schema should be updated to specify the allowable data types for all request records and API firewalls can be used to enforce this contract and eliminate any injection of additional parameters. This topic will be explored fully in *Chapter 11, Shield Right for APIs with Runtime Protection*.

Implementation vulnerabilities

In the final portion of this chapter, we will cover the remaining vulnerabilities without going into a deep level of detail. Many of these vulnerabilities are not specific to APIs but affect many software systems, and there is already a body of knowledge in the industry on how to address these issues. I have included a *Further reading* section at the end of this chapter, and as always, the reader is well advised to consult the OWASP project, which maintains excellent guides and resources on these topics.

Injection

Injection vulnerabilities have plagued software systems for over two decades already; fortunately, it is a vulnerability class that can be totally eliminated by implementing patterns for secure coding. Injection attacks occur when a system trusts user input without validating it, allowing an attacker to launch attacks against underlying components such as databases, operating systems, and filesystems. Typical examples include SQL injection, command injection, and XSS.

The top recommendation to protect against injection is to validate (and never trust) user input fully. Many well-developed and tested validation libraries for various languages and frameworks give API developers a reliable way to protect their APIs from untrusted input data.

For SQL injection, the underlying database queries should always use the parameterized format of the database queries in preference of the wildcard queries since the latter is vulnerable to being manipulated by malformed input data. By using parameterized queries, the database layer ensures that injection is prevented.

Use filters to sanitize all user input data and ensure that any special characters (such as path characters or SQL syntax characters) are properly handled by either escaping safely or rejecting them entirely.

Using OpenAPI definitions, the format of all data inputs can be fully constrained, and API firewalls can be used to protect the API before any invalid input data reaches the API itself. This topic will be explored fully in *Chapter 11, Shield Right for APIs with Runtime Protection*.

Server-Side Request Forgery

Server-Side Request Forgery (**SSRF**) is a category of flaw that has been prevalent for some decades but has recently attracted notoriety in the API world due to its inclusion in the OWASP API Security Top 10 2023. In this attack, an attacker can force the server to redirect to a URL without validating the URL. This allows an attacker to redirect to potentially dangerous websites, or to load resources from a malicious source.

The easiest way to protect against this attack vector (and the one recommended here) is to use an explicit *allow list* of allowed URLs open for redirection. This does require prior knowledge of the API and the likely range of expected redirect URLs.

Failing that, the following steps should be taken to safeguard against harmful redirects:

- Restrict the URL schema and ports allowed
- Disable HTTP redirections
- Do not send raw responses to the client
- Restrict the allowable media types

SSRF is one of the more difficult flaws to protect against since being too restrictive in the allow list (in order to achieve optimum security) runs the risk of degrading application functionality. This is one of the vulnerabilities where a trade-off between security and functionality is required.

Insufficient logging and monitoring

This is a category where there is no specific recommendation relating to API security. As with any software system, it is imperative that sufficient logging and monitoring be provisioned to monitor the following:

- Normal operation of the API (there is not an excess of errors or failures)

- Failed operations are recorded including all transaction details to trace the root cause of the failed operation

- Suspicious transactions are identified, and appropriate alerts or alarms are raised

The most important consideration in the logging and monitoring of APIs is to ensure that the logging and reporting are done in a manner consistent with the existing organizational standards. Where possible, it allows API logs to be ingested into standard SIEM and SOC solutions from where further monitoring and analysis can be performed. Resist the temptation to perform too much analysis locally, where you may need a complete picture of the overall risk and threat landscape.

Avoid logging sensitive data such as PII and credential information, as these logs may persist and be accessible by persons not entitled to them.

Protecting against unrestricted resource consumption

The primary way to protect against the overuse of API resources is to implement rate limiting and throttling on your APIs. API rate limiting monitors the access to an API endpoint for a given client (usually based on IP address) and checks to see whether a predetermined allowed number of accesses has been made within a given window. If so, then the client will be rate-limited, typically with **429 Too Many Requests**. The client will have the option to back off and retry the request or fail outright.

The server uses several different algorithms to detect the rate-limiting threshold, and some may be quite adaptive to only trigger in extreme cases of abuse. For example, the server can block many requests over a wide window or may only block on very high peak demands (or bursts) of access. The choice will depend on the perceived threats to the API, for example, denial-of-service attacks or mass data exfiltration.

Rate limiting can be implemented in the API code itself using standard libraries for different languages and frameworks. This method is somewhat fragile since it means that the API will still have to process each request to determine if it exceeds the threshold limit – just this amount of processing can cause the API to become overloaded.

The more robust and preferred option is to use rate limiting in either an API gateway or an API firewall, where the processing burden can be offloaded to dedicated processors. This topic will be explored fully in *Chapter 11, Shield Right for APIs with Runtime Protection*.

Finally, the API server should also be designed in such a way as to allow for scaling of resources to gracefully cope with variable loads that may become higher than expected at times. Modern cloud-native platforms allow for seamless, automatic scaling, and API platforms should be designed to avail themselves of such features to avoid failure of services.

API designers should avoid situations where they return excessive amounts of data that may cause memory exhaustion and ensure that data is paginated into manageable page sizes.

Rate limiting versus throttling versus quotas

There is sometimes confusion regarding the different types of limits placed on API access under different circumstances. Let us disambiguate the common terms:

Rate limiting: In this case, the API enforces a hard limit (stops all further transactions for a given period) due to the API client's excessive use of the API in the preceding time period. Typically, the number of transactions in a window is measured for a particular client (usually identified by IP address). If it exceeds a threshold, then further access is blocked for a period. The API typically returns a **429 Too Many Requests** status code and may include additional X headers to reflect the current rate-limiting status.

Throttling: This technique prevents overload of a server or API resource by limiting the number of requests that can be processed and is typically achieved by taking the server offline or buffering the transactions (by introducing a delay). Throttling limits at the server level rather than at the calling client level.

Quotas: While the two previous techniques are mainly security or availability controls, quotas are used to implement controls based on the business model of the API. API access is commonly controlled based on a subscription model (often with a free but restricted model and paid options), and quotas dictate how many calls can be made within a particular billing window. Quotas have no bearing on API security at all; they are purely a billing artifact.

Defending against API business-level attacks

Finally, let us conclude this chapter by looking at the two business-level vulnerabilities included in the *OWASP API Security Top 10 2023*.

Unrestricted access to sensitive business flows

This vulnerability results when an attacker can abuse the standard flow of an API-based application to subvert the original business intent to their benefit. Typical examples include abuse of airline ticket booking systems, online event ticketing systems, or various online retailers.

The exact nature of the abuse will depend on the specifics of the industry and how their API design maps to their business flows. Unfortunately, this is one of the hardest vulnerabilities to defend against because, unlike many of the others covered already, there is no single point of fix. In fact, the APIs themselves may be flawless, but by being used in a nefarious fashion, they expose the business to risk.

Typically, risk minimization involves both the business and engineering teams. Business owners can refactor how they offer their products and services to reduce the likelihood of abuse (for example, by restricting tickets available for sale, or prohibiting the resale of items), whilst engineering should consider the following techniques:

- Ensure that devices are identified and in the case of scrapers or headless devices, are barred from access

- Make efforts to verify human identity, for example by using MFA step-up authorization at key points in the flow

- Identify patterns of non-human behavior, for example, impossibly fast transactions

Unfortunately, this is one of the more challenging threat vectors for API-based businesses and one of the rising threats.

Unsafe consumption of APIs

APIs facilitate the connectivity of businesses providing services to customers and consumers while consuming upstream API providers to deliver business value. Although upstream APIs may be considered an integral component in the supply chain, they are increasingly a risk vector for organizations.

Upstream APIs may be considered trusted, but they may expose their unwary consumers to risk. The advice on injection attacks never to trust external input is equally valid when consuming third-party APIs – never place implicit trust in the output from an API. Always validate the input and apply appropriate filtering techniques. Also, always ensure that connectivity to third parties is via a secured transport and uses enhanced authentication such as mutual certificates.

Finally, as a consumer of APIs, due diligence should be conducted into the security posture of upstream providers. For example, do they provide evidence of API security testing or other indications of proactive security measures, such as developer training?

Summary

This has been a key chapter in your journey in learning how to defend APIs from attacks. The key learning point is that while the attack vectors are vast and varied, there are well-established patterns for defending APIs against common vulnerabilities.

Firstly, we examined how to deal with common authentication vulnerabilities, focusing on best practices for handling JWTs securely, using OAuth2 securely, and securing your passwords, tokens, and reset process. Whilst the recommendations are extensive, developers will benefit from following these guidelines to avoid some of the most nefarious authentication vulnerabilities. Secondly, authorization of your APIs poses significant challenges, and in this chapter, we learned how to address both object-level and function-level vulnerabilities through the judicious usage of authorization middleware to bolster your defenses.

Thirdly, we dealt with the critical vulnerabilities associated with API data, namely excessive data exposure and mass assignment. We learned how these vulnerabilities can be comprehensively addressed using defensive coding patterns when handling data objects. Finally, we covered several miscellaneous topics, such as injection attacks, SSRF attacks, unrestricted resource consumption, and business-level attacks.

In the next chapter, we will look at how to securely implement APIs using various languages and frameworks.

Further reading

These links provide further reading on handling JWTs:

- `https://42crunch.com/7-ways-to-avoid-jwt-pitfalls/`
- `https://auth0.com/blog/a-look-at-the-latest-draft-for-jwt-bcp/`
- `https://curity.io/resources/learn/jwt-best-practices/`
- `https://redis.com/blog/json-web-tokens-jwt-are-dangerous-for-user-sessions/`

These links provide further reading on implementing OAuth2:

- `https://pragmaticwebsecurity.com/files/cheatsheets/oauth2securityfordevelopers.pdf`
- `https://datatracker.ietf.org/doc/html/draft-ietf-oauth-security-topics`
- `https://auth0.com/resources/ebooks/oauth-openid-connect-professional-guide/`
- `https://www.moesif.com/blog/technical/cors/Authoritative-Guide-to-CORS-Cross-Origin-Resource-Sharing-for-REST-APIs/`
- `https://curity.io/resources/learn/spa-best-practices/`
- `https://datatracker.ietf.org/doc/html/rfc6819`

These links provide further reading on using authorization middleware:

- `https://casbin.org/`
- `https://www.osohq.com/`
- `https://www.osohq.com/post/why-authorization-is-hard`

These links provide further reading on authorization protection:

- `https://www.hivesystems.io/blog/are-your-passwords-in-the-green`
- `https://inonst.medium.com/a-deep-dive-on-the-most-critical-api-vulnerability-bola-1342224ec3f2`

These links provide further reading on data vulnerabilities:

- `https://fastapi.tiangolo.com/tutorial/extra-models/`
- `https://cheatsheetseries.owasp.org/cheatsheets/Mass_Assignment_Cheat_Sheet.html`
- `https://dev.to/izabelakowal/some-ideas-on-how-to-implement-dtos-in-python-be3`

These links provide further reading on miscellaneous vulnerabilities:

- `https://owasp.org/www-community/Injection_Flaws`
- `https://cheatsheetseries.owasp.org/cheatsheets/Server_Side_Request_Forgery_Prevention_Cheat_Sheet.html`
- `https://cheatsheetseries.owasp.org/cheatsheets/Server_Side_Request_Forgery_Prevention_Cheat_Sheet.html`
- `https://nordicapis.com/everything-you-need-to-know-about-api-rate-limiting/`

10

Securing Your Frameworks and Languages

In this chapter, we will discuss the core activity involved in converting an API design or specification into working code. This is where the rubber meets the road for API security—while having a well-designed API with security considered is essential, the running API will only be as secure as the implementation allows. As an API developer, it is essential that you understand the best practices for coding your API securely and using your frameworks and libraries in a secure manner.

Firstly, we will focus on the challenges of managing a design-first process, where you will learn how to manage the lifecycle of both your API specification and the underlying code implementing the API. Generating API code from the API specification automatically is a tenet of a secure design-first process, and you will learn how to use modern code-generation tools. Finally, we will examine best practices in your frameworks to ensure secure API implementations.

In a nutshell, this chapter is going to cover the following main topics:

- Managing the design-first process in the real world
- Using code-generation tools
- Using OpenAPI with frameworks
- Patterns for securing frameworks

Technical requirements

For this chapter, you will need a development machine capable of the following:

- Running Docker locally
- Running VS Code with various marketplace extensions
- Access to the internet and a GitHub account to access the examples

This chapter contains many code samples in various languages; these can either be run locally, which will require the installation of compilers, SDKs, and frameworks, or from within a Docker build container.

The example code and various breaking changes to the instructions can be found in the `Chapter 10` folder on the book's GitHub repository here: `https://github.com/PacktPublishing/Defending-APIs/tree/main/Chapter10`

Managing the design-first process in the real world

In *Chapter 8, Shift-Left for API Security,* we examined the concepts associated with design-first API development, namely that the API development team starts with the API design first (via an OpenAPI definition) and then proceeds to implement the API code. There are numerous benefits to adopting this approach, including the following:

- **Incorporating security early**: Incorporating security early in the design lifecycle ensures that designers and developers are forced to consider how they will secure their APIs (for example, an OAS definition can be parsed to check if a security method has been specified and generate a warning if not). This makes it harder to leave security considerations for a later stage in the lifecycle.

- **Automated document generation**: A well-formed OAS definition can be used to generate comprehensive API documentation for an API developer portal, making it easier for consumers to integrate with the API.

- **Automated generation of mocks and stubs**: Since the API request and response data are defined, the OAS definition can be used to generate mocks and stubs, which can mimic the actual behavior of the API implementation.

- **Automated security testing**: Since the OAS definition specifies the security constructs, automated security testing can be performed using tools such as 42Crunch's API security scanner.

- **Generation of test code**: The OAS definition defines the expected API responses under certain conditions, and test frameworks can use this information to generate automatic tests (such as in Postman).

- **Generation of client and server code**: Finally, the most useful thing is the ability to generate client and server code stubs based on the OAS definitions. We will explore this feature in the next section, *Using code-generation tools.*

Despite the numerous significant advantages of using a design-first approach, the situation in the real world is a little more nuanced. Traditionally, APIs have been developed using a **code-first** approach where developers would capture business requirements and implement the API directly in code, which becomes the source of truth for the API. Any changes required to be made to the API are made directly in code, ideally using a change management process with a source code management system. The question that arises is where the OAS definition fits into this process. If the OAS definition is not required as part of the design or implementation, is there any need for it at all? The answer is often that a definition is needed, often for documentation purposes (as an API consumer, it is easier to consume an API with an OAS definition than an API without one).

The solution to this apparent lack of OAS definition in the code-first approach is to use software tools and techniques (such as introspection and reflection) to reverse-engineer a definition from the running code. We will investigate such tools for each language as we progress through the chapter.

This flow is illustrated in the bottom half of the cycle in *Figure 10.1*. In this cycle, the OAS definition is obtained from the implemented code and is typically used for documentation purposes, such as the well-known Swagger UI interface (for example, see `https://petstore.swagger.io/`):

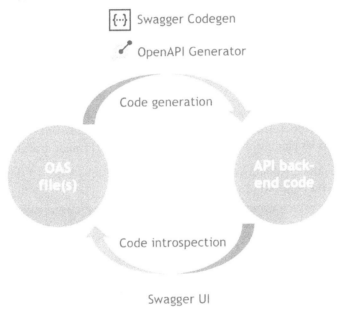

Figure 10.1 – OAS generation versus introspection lifecycle

While this illustrates that it is possible to reverse-engineer an OAS definition from a code-first implementation, I would add that this is not the preferred approach for a modern API development process. The benefits of design-first far outweigh the additional effort of the upfront design process, particularly in enforcing a shift-left security approach. To the extent possible, organizations should adopt the flow in the top half of *Figure 10.1* to benefit from a design-first approach.

In reality, you will encounter a mix of the two approaches, and the burning question is how to manage this process as the design iterates. Where is the source of truth, and how do we ensure that changes are made at the right place in the lifecycle? For example, in a code-first scenario, there is no point in making changes to the OAS definition to incorporate design changes since the code is the source of truth.

In this book, I favor the API design-first approach since it allows security to be considered as early as possible in the lifecycle, and having an OAS definition has a number of ancillary benefits as outlined previously.

Using code-generation tools

In this section, we will deal with the practicalities of generating client and server code from an OAS definition using the two most popular tools currently in use:

- Swagger Codegen (`https://swagger.io/tools/swagger-codegen/`)
- OpenAPI Generator (`https://swagger.io/tools/swagger-codegen/`)

Swagger Codegen

Swagger **Codegen** (`https://swagger.io/tools/swagger-codegen/`) is part of the SmartBear API design and test tool suite. The product can be used within the **SwaggerHub** web portal or as a standalone command-line tool. The **SwaggerHub** portal offers a limited-time free trial, after which you can upgrade to various paid plans, but the command-line tool is made available free of charge.

Firstly, let us examine the capabilities of the **SwaggerHub** portal. *Figure 10.2* shows a small OAS definition loaded into the editor window:

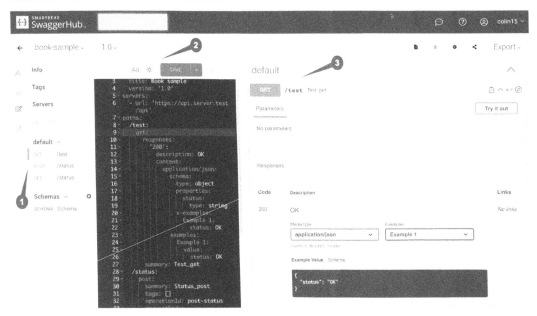

Figure 10.2 – Swagger Codegen user interface

The following areas are available to the API designer:

1. The tree view of the OAS definition shows the endpoint names and operations

2. The OAS definition in source code editing mode allows changes to be made in real time

3. A Swagger UI window allows the user to interact with the API by submitting requests to the API (assuming a server instance is operational)

From this view, it is possible to generate both client and server code using the **Export** button in the top-right of the screen, shown in *Figure 10.3*:

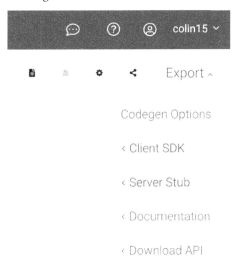

Figure 10.3 – Export functions on SwaggerHub

For illustration, let us generate a Python Flash server stub for this API. Firstly, the code-generation options can be set to override some of the package options, such as the server port and the support for Python2. This is shown in *Figure 10.4*:

Figure 10.4 – Code-generation options on SwaggerHub

All that's left to do is to click the link to generate a server stub for Python Flask and a ZIP archive will be generated, which can be downloaded and extracted locally. An example of the aforementioned OAS definition is shown in *Figure 10.5*:

Figure 10.5 – Package contents of Python Flask API server stub (swagger-codegen)

Figure 10.5 shows that the SwaggerHub platform has generated a fully featured Python package, including an application with initialization code, controllers, and data models, and a set of dependencies in `requirements.txt`. The package also includes a Dockerfile, and it embeds the associated `swagger.yaml` into the solution. The startup code shown adds the Swagger UI to the application at the `/swagger` URL. The custom code within the data models and controllers must be added to the API, but a tremendous amount of low-value boilerplate code has been automated by the SwaggerHub platform.

The same generator can be run at the command line using the `swagger-codegen` CLI:

```
colind@mbm1pro: ~/ #
swagger-codegen generate -i book_sample_1.yml -l python-flask -o ./
out/server/
```

OpenAPI Generator

While the SwaggerHub generator is powerful and quite well-featured, its roadmap, features, and bug fixes are at the behest of the SwaggerHub development team, who may have more pressing demands. To address these concerns, the open source community has stepped in to create an open source implementation of an API code generator based on the 2.4.0-SNAPSHOT of the Swagger Codegen repository. This is the **OpenAPI Generator** project, which we will examine in this section (https://openapi-generator.tech/).

According to the FAQ on the OpenAPI Generator website, the founding members felt that "*having a community-driven version allows for innovation, reliability, and a roadmap owned by the community*".

The OpenAPI Generator can be installed using several methods, including via **npm**, **Homebrew** on MacOS, or **Scoop** on Windows; as a JAR file for any JVM platform; via a Docker image; or using various plugins such as **Maven** and **Gradle**. I will use the MacOS CLI or the Docker image in the following sections.

Generating server stubs

To generate a server stub, the same command format as swagger-codegen is used, as shown in this example:

```
colind@mbm1pro: ~ # openapi-generator generate -i book_sample_1.yml -g
python-flask -o out/
```

Similar to the Swagger Codegen example, a full-fledged Python project is generated, as shown in the following figure:

Figure 10.6 – Package contents of Python Flask API server stub (openapi-generator)

If you are familiar with Swagger Codegen, then things will seem familiar so far. Let us now look at some of the interesting new features OpenAPI Generator provides.

Generating schema

In order to make efficient use of security tooling in production, such as API firewalls and gateways, you should define and maintain your schema in your OAS definition (for example, for the enforcement of API firewall rules or gateway data inspection). However, since the same data schema will be needed to create the backend databases, there is a danger that the same data definitions could exist in two locations: in the API definition and the database model. Ideally, this problem would be solved by having a single source of truth, which would be the OAS definition according to the API design-first philosophy.

Fortunately for us, the OpenAPI Generator provides us with the option to generate a variety of database schema from the data definitions within an OAS definition. This is best illustrated with an example of an API with a well-specified `user` entity (certain details have been omitted for clarity):

```
"UsersItem": {
    "required": [
        "email", "password", "name", "_id", "is_admin", "account_
balance", "onboarding_date"
],
    "properties": {
     "_id": {   "format": "uuid",
             "maxLength": 36,
             "type": "string"   },
     "email": {   "maxLength": 32,
             "type": "string" },
     "password": {   "maxLength": 64,
             "type": "string" },
     "name": { "maxLength": 50,
             "type": "string" },
     "is_admin": {   "type": "boolean",
             "description": "is admin" },
     "account_balance": { "maximum": 1000,
             "type": "number",
             "format": "float" },
      "onboarding_date": { "maxLength": 38,
             "type": "string",
             "format": "date-time" }
 } }
```

The highlights indicate particular constraints on the field types, such as string length, numeric types, and a timestamp field. The OpenAPI Generator can be run against this definition using the `mysql-schema` output type, as shown here:

```
colind@mbm1pro: ~ # openapi-generator generate -i Pixi.json -g mysql-schema -o out/
```

This produces the following MySQL script for this particular data type:

```
CREATE TABLE IF NOT EXISTS `UsersItem` (
    `_id` CHAR(36) NOT NULL,
    `email` TEXT NOT NULL,
    `password` VARCHAR(64) NOT NULL,
    `name` VARCHAR(50) NOT NULL,
    `is_admin` TINYINT(1) NOT NULL COMMENT 'is admin',
    `account_balance` DECIMAL(20, 9) NOT NULL COMMENT 'remaining balance',
    `onboarding_date` DATETIME NOT NULL
) ENGINE=InnoDB DEFAULT CHARSET=utf8mb4 COLLATE=utf8mb4_unicode_ci;
```

This SQL script shows how well the OAS definition has been translated into the corresponding database layer. The string lengths have been translated to the SQL syntax, the `boolean` type has been translated to a `tinyint` type, the `float` has been translated to a `DECIMAL(20, 9)`, and the timestamp field has been translated to a `DATETIME` field.

This simple example illustrates how easily the OpenAPI Generator can translate a data type from the OAS domain to the database domain without any manual coding.

Generating documentation

The OpenAPI Generator also excels in producing human-readable documentation from the OAS definition in a variety of formats. The following command produces HTML documentation:

```
colind@mbm1pro: ~ # openapi-generator generate -i Pixi.json -g html -o out/
```

Figure 10.7 shows the top-level API summary documentation containing the API preamble and the summary of API calls with hyperlinks to the individual methods:

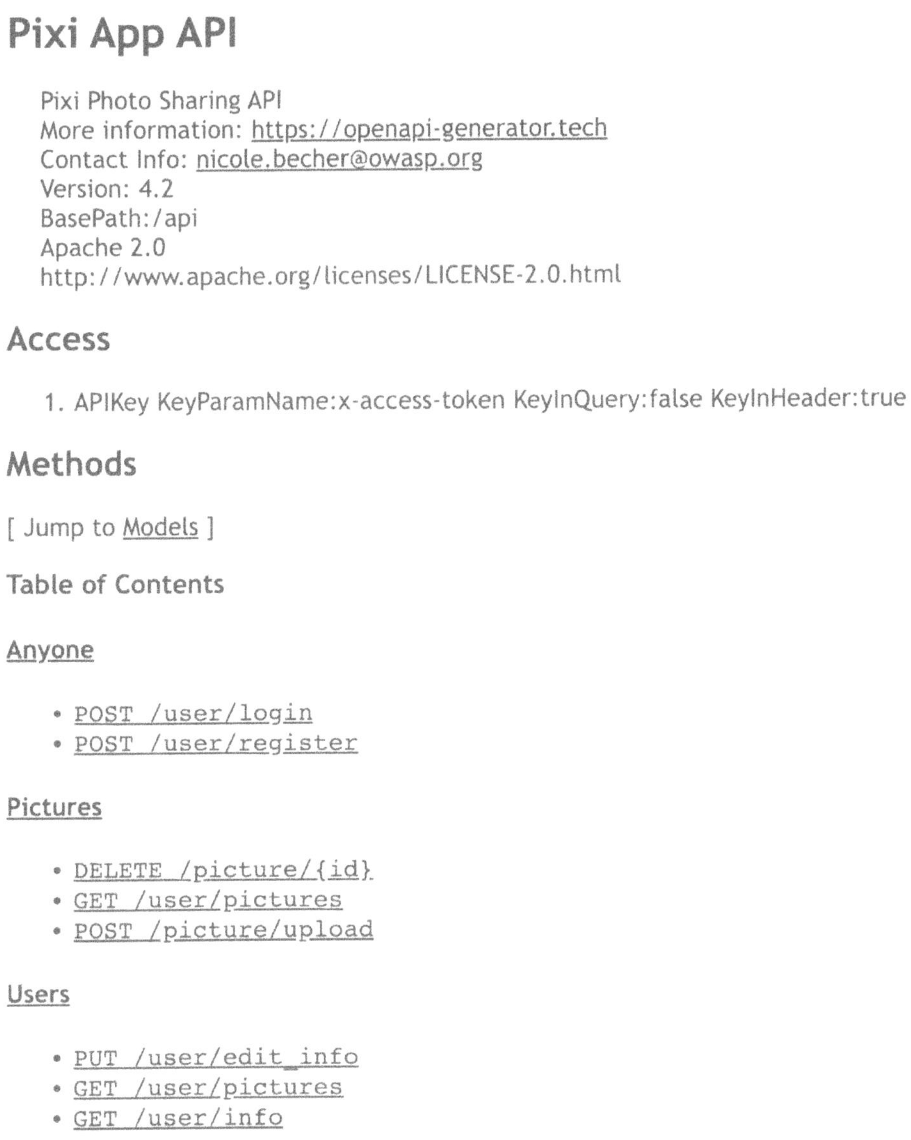

Pixi App API

Pixi Photo Sharing API
More information: https://openapi-generator.tech
Contact Info: nicole.becher@owasp.org
Version: 4.2
BasePath:/api
Apache 2.0
http://www.apache.org/licenses/LICENSE-2.0.html

Access

1. APIKey KeyParamName:x-access-token KeyInQuery:false KeyInHeader:true

Methods

[Jump to Models]

Table of Contents

Anyone

- POST /user/login
- POST /user/register

Pictures

- DELETE /picture/{id}
- GET /user/pictures
- POST /picture/upload

Users

- PUT /user/edit_info
- GET /user/pictures
- GET /user/info

Figure 10.7 – HTML documentation for top-level API

Figure 10.8 shows the detailed documentation for an individual API method, in this case, the /user/login method:

POST /user/login   Up

user/password based login (**authenticate**)

user supplies user name and password and receives a json web token

Consumes
This API call consumes the following media types via the Content-Type request header:

   • application/json

Request body
   authenticate_request <u>authenticate_request</u> (required)
   Body Parameter

Return type
<u>authenticate_200_response</u>

Example data
Content-Type: application/json

```
{
  "_id" : "a83a29f5-0d63-46f2-8f2e-44c2f1d2e07e",
  "message" : "message",
  "token" : "token"
}
```

Produces
This API call produces the following media types according to the Accept request header; the media type will be conveyed by the Content-Type response header.

   • application/json

Responses
200
<u>authenticate_200_response</u>
422
missing parameters <u>ErrorMessage</u>
default
unexpected error <u>ErrorMessage</u>

Figure 10.8 – HTML documentation for individual methods

Figure 10.9 shows a Markdown view of the previous `UserItem` used in the schema generation examples:

UsersItem

Properties

Name	Type	Description	Notes
_id	UUID		[default to null]
email	String		[default to null]
password	String		[default to null]
name	String		[default to null]
is_admin	Boolean	is admin	[default to null]
account_balance	Float	remaining balance	[default to null]
onboarding_date	Date		[default to null]

[Back to Model list] [Back to API list] [Back to README]

Figure 10.9 – Markdown documentation for API data types

Although it's not essential to API security, having well-documented APIs is vital for the successful adoption of your APIs, and OpenAPI Generator makes documentation generation a simple matter.

Using templates and custom generators

One of the most powerful features of the OpenAPI Generator is the sensible built-in behaviors supporting various server and client languages and API configurations. If the built-in behavior does not match your requirements, the generator behavior can be customized in several ways, or you can build your own language support entirely from scratch and contribute it back to the project.

While researching this chapter, I encountered an issue where there was a hyphen in the name of an API variable, which was causing the Python execution to fail (hyphens are forbidden in type names in Python). To debug this, I used the template feature to extract the **Mustache** (a common web templating language) templates for the `python-flask` generator as follows:

```
colind@mbm1pro: ~ # openapi-generator author template -g python-flask
```

This command dumps the associated templates into an output folder, as shown in *Figure 10.10*:

```
~ __init__.mustache                        {{/isApiKey}}
~ __init__model.mustache                   {{#isBasicBasic}}
~ __init__test.mustache
~ __main__.mustache              55    def info_from_{{name}}(username, password, required_scopes):
~ base_model.mustache                      """
~ controller_test.mustac...                Check and retrieve authentication information from basic auth.
~ controller.mustache                      Returned value will be passed in 'token_info' parameter of your operation function, if there is
~ Dockerfile.mustache                      one.
~ dockerignore.mustache                    'sub' or 'uid' will be set in 'user' parameter of your operation function, if there is one.
~ encoder.mustache
~ git_push.sh.mustache                     :param username login provided by Authorization header
~ gitignore.mustache                       :type username: str
~ model.mustache                           :param password password provided by Authorization header
~ openapi.mustache                         :type password: str
~ param_type.mustache                      :param required_scopes Always None. Used for other authentication method
~ README.mustache                          :type required_scopes: None
~ requirements.mustache                    :return: Information attached to user or None if credentials are invalid or does not allow
~ security_controller.mu...                access to called API
~ setup.mustache                           :rtype: dict | None
                                           """
                                           return {'uid': 'user_id'}
```

Figure 10.10 – Mustache template for the python-flash security controller

There is a template for each of the project components (right down to the Dockerfile and README file), and to understand better how the template works, let us look at the `security_controller` template highlighted in *Figure 10.10*. For a Mustache template, anything within a `{{ }}` demarcation has a special meaning; any other content is emitted verbatim to the output file.

The security controller has multiple functions that can generate a security controller function based on the API's specified security type (more on this in the next section). The conditional logic at the top of the function will emit a handler for basic authentication if `isApiKey` is false and `isBasicBasic` is true, as shown:

```
{{/isApiKey}}
{{#isBasicBasic}}
```

The function emitted takes a `username` and `password` as expected for basic authentication. The function's body is empty except for some comments and a return statement; it is necessary to implement the functionality in a working solution.

In my scenario, the issue was the name variable containing a hyphen, which caused the generated Python code to be invalid. I tweaked the template to suppress the hyphen, and the template-generated code then executed without issue. Although this is a slightly contrived example, it shows how the template engine works and how easy it is to modify it to suit your scenario—for example, the template could contain actual code for a basic authentication logic handler for your application.

If modifying the existing templates is still insufficient for your needs, the OpenAPI Generator allows you to write your custom generator from scratch by providing you with the boilerplate for a new generator. I found the customization features of OpenAPI Generator extremely powerful; it has been designed to be extensible and future-proof.

Configuration of OpenAPI Generator

Finally, let us look at some of the configuration options for OpenAPI Generator, particularly regarding security support.

Although the OpenAPI Generator does a good job of trying to harmonize support for features across all client and server types, there are variations in support for more esoteric features. For example, when using `python-flask`, many of the more common OAuth2 flows are unsupported, whereas on `java-micronaut-server`, all OAuth2 flows are supported. The reader is advised to check the individual generator support page to determine specific feature support.

Global properties

Global features allow the user to specify properties such as the host name, the base path, various information fields, supported schemes, documentation, and examples.

Data and schema support features

The data types specify which data types the generator supports, from simple data types such as integers and strings to complex data types such as collections and arrays. The schema support specifies which of the composition features are supported, such as simple and composite, and the associations, such as `allOf, anyOf, oneOf`.

Transport format features

The transport format features specify the format of the data transferred between the client and API server and can be JSON, XML, ProtoBuf, or Custom. Nearly all generators support the JSON and XML formats.

Security features

The security features configuration specifies which of the OpenAPI specification security schema will be used for the API. If you need a reminder on some of these topics, please refer to *Chapter 2, Understanding APIs*, specifically the section on *Access control*.

Table 10.1 shows the different secure features and their descriptions:

Feature name	Description
BasicAuth	Specifies the use of HTTP basic authentication (username/password)
ApiKey	Specifies the use of an API key (either in query, header, or cookie)
OpenIDConnect	Uses the OpenIDConnect standard for authentication
BearerToken	Specifies the use of a bearer token, usually JWTs
OAuth2_Implicit	Specifies the OAuth Implicit Flow
OAuth2_Password	Specifies the OAuth Resource Owner Password Flow
OAuth2_ClientCredentials	Specifies the OAuth Client Credentials Flow
OAuth2_AuthorizationCode	Specifies the OAuth Authorization Code Flow
SignatureAuth	No details are available; I suggest not using it

Table 10.1 – OpenAPI Generator security features

OpenAPI Generator will parse the OAS definition, detect security parameters, and generate the corresponding code. Consider a simple OAS fragment showing a GET endpoint named /status that requires basic authentication:

```
/status:
  get:
    summary: Status_get
    operationId: get-status
    responses:
      '200':
        description: OK
    security:
      - BasicAuth: []
```

Using java-micronaut-server as the generator, OpenAPI Generator generates the template code shown in *Figure 10.11* for the GET method for /status:

```
@Generated(value="org.openapitools.codegen.languages.JavaMicronautServerCodegen", date="2023-10-03T13:37:05.259057+01:00[Europe/London]")
@Controller
@Tag(name = "Default", description = "The Default API")
public class DefaultController {
    /**
     * Status_get
     *
     */
    @Operation(
        operationId = "getStatus",
        summary = "Status_get",
        responses = {
            @ApiResponse(responseCode = "200", description = "OK")
        },
        security = {
            @SecurityRequirement(name = "BasicAuth")
        }
    )
    @Get(uri="/status")
    @Produces(value = {})
    @Secured({SecurityRule.IS_AUTHENTICATED})
    public Mono<Void> getStatus() {
        // TODO implement getStatus();
        return Mono.error(new HttpStatusException(HttpStatus.NOT_IMPLEMENTED, null));
```

Figure 10.11 – GET method with basic authentication

The code shows several security annotations decorating the `getStatus()` method:

- `security = {@SecurityRequirement(name = "BasicAuth")`

- `@Secured({SecurityRule.IS_AUTHENTICATED})`

All that remains for the developer to do is to implement the business logic of the `getStatus()` method. Using a code generator such as OpenAPI Generator greatly reduces the risk of human errors, such as omitting a security decorator, which would leave a method unauthenticated.

Using OpenAPI with frameworks

Figure 10.1 at the start of this chapter perfectly illustrates the lifecycle of the OAS definition within a typical SDLC. Ideally, the top path (labeled *Code Generation*) goes from an OAS definition to API code courtesy of a code generator tool, as described in the preceding section. However, what about the bottom path (labelled *Code Introspection*), where API code already exists without an OAS definition? This will generally depend on the language and framework you have used for your API development. Typically, the approach involves loading a library with your API application, which will use introspection and reflection techniques to discover methods that provide API functionality. Often, these methods will be decorated with various annotations that aid the introspection process (for example, look at the code in *Figure 10.11* to see examples of such annotations).

While this general approach applies to all languages and frameworks, there are some specific recommendations or considerations. Let us look at some of the most common scenarios.

For **Java**, reverse-engineering API code is generally well supported, either directly in the framework or with plugins. The most popular framework for Java APIs and microservices is the Spring Boot framework, and for this, the de facto OpenAPI support library is **Springdoc-OpenAPI** (`https://springdoc.org/`). This library supports many options, such as advanced UI configuration and support for error decorators. The other main contender of APIs and microservices is **Micronaut**, and here, comprehensive support is built in out of the box (`https://micronaut-projects.github.io/micronaut-openapi/latest/guide/`).

If you are using **.NET Core**, then you are well supported by the ability to choose between two equally popular and capable packages: **NSwag** (`https://github.com/RicoSuter/NSwag/wiki`) and **Swashbuckle** (`https://github.com/domaindrivendev/Swashbuckle.AspNetCore`). They offer similar features, including the ability to produce OAS definitions from existing .NET Core code bases (even in compiled form). NSwag offers some code-generation capabilities (similar to OpenAPI Generator) and even features a GUI design studio.

Python is a mixed bag, requiring the user to install one or more support packages and do a certain amount of manual configuration. The most popular package at the core of most solutions is **apisec** (`https://github.com/marshmallow-code/apispec`). Depending on the particular frameworks used (such as **Flask** or **aiohttp**), installing other support packages as documented on the **apisec** GitHub page might be necessary.

Finally, for **Node.js**, the situation is similar to Python, with **swagger-ui-express** (`https://www.npmjs.com/package/swagger-ui-express`) being the de facto support library. From the documentation, this requires an OAS definition since it cannot use introspection to reverse-engineer the OAS definition.

Code examples for languages and frameworks

I have only identified a few common languages and frameworks; there is a much wider selection available to API developers. Since OAS support is a rapidly evolving topic, the reader should refer to the book's accompanying website, where I will provide code samples for the frameworks covered and many others popular in API development.

Managing the lifecycle of the OAS definition and the code base

Figure 10.1 illustrated the "chicken or egg" nature of the OAS definition and its corresponding codebase. In a real-world scenario, it is necessary to maintain the OAS definition (as input to your API gateway, for example), and the code base should match the intended business logic. How are the two to be kept in sync?

In the first scenario shown in *Figure 10.12*, the OAS definition and the codebase track each other very closely. Sometimes the code comes first, and other times the OAS definition is updated first. The two are kept up to date via manual reviews and patching directly based on changes. As long as the deltas are small, this process is quite manageable.

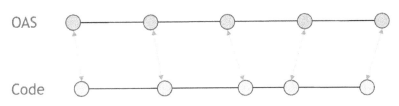

Figure 10.12 – Continuous OAS definition and codebase development

The second scenario, shown in *Figure 10.13*, is probably the most common in the real world, and that is where the codebase changes frequently according to feature development. Occasionally, the OAS definition is extracted from the running codebase (using one of the code introspection methods described in *Using OpenAPI with frameworks*). This ensures that an accurate OAS definition is obtained from the implemented code for deployment into API gateways and other tools. The reverse-engineering step is typically correlated with major releases to ensure the releases are up to date in terms of both the code and the OAS definition.

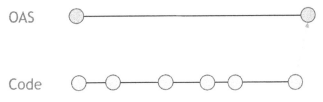

Figure 10.13 – Code-first development

The third scenario in *Figure 10.14* is the least common by some margin, and here, the codebase (or at least parts of it, such as the endpoint stubs and authorization and authentication handlers) is regenerated using a tool such as OpenAPI Generator, described previously. This method is typically only useful with a stable feature set, where little new code is being developed and only infrequent changes are made to the OAS definition or codebase.

Figure 10.14 – OAS definition-first development

Patterns for securing frameworks

Finally, this chapter concludes with some general advice on securing your API frameworks to produce more robust and secure APIs. The advice is not specific to particular API vulnerability types (for a deep dive on this, refer to *Chapter 9*, *Defending against Common Vulnerabilities*) but rather focuses on securing the API runtime to serve APIs securely.

Secure defaults

Not all frameworks are created equally, which is particularly pertinent to the framework's security. The best recommendation is to ensure that you understand the default settings of your framework, and that includes (but is not limited to) the following:

- Is HTTP enabled by default? Preferably, disable HTTP and use HTTPS.

- Are any administrative or diagnostic interfaces enabled by default? Disable all but the essential interfaces.

- What HTTP methods are enabled by default? For instance, many frameworks will respond to a DELETE operation with the default handler.

- Is any middleware enabled by default? Configure middleware correctly to authenticate and authorize your APIs.

- Is anonymous access enabled by default? Unless specifically required, disable all anonymous access.

- What is the behavior if a non-existent endpoint is accessed? Ensure that your framework does not default to dangerous behavior and rather just ignores an invalid request.

- How does the framework behave if the supplied data differs in type from that specified in the content headers? Test to verify that the framework does not produce unexpected results with invalid data input.

- How does the framework behave under excessive load or with excessively large data payloads? Does it reject them safely or attempt to process them?

Start by researching your framework's default configuration and behaviors and consult online resources (such as this book's GitHub repository) for guidance on hardening the framework. Ensure you enforce this hardened configuration using your Infrastructure-as-Code tool of choice (Terraform, Helm, Ansible, etc.). A basic step such as this will greatly reduce residual risk in your framework and requires modest effort.

Using your middleware

One of the major drawbacks of frameworks is that they are complex and hard to understand; one of the major benefits of frameworks is that they are comprehensive and provide a lot of functionality out of the box. Take the time to understand the features provided by the middleware of your framework and learn patterns for robust implementations of common scenarios—let the middleware do the heavy lifting and avoid reinventing the wheel.

As an example of a well-implemented middleware stack, let us look at some of the capabilities provided by the ASP .NET Core framework:

- Authentication (from simple basic authentication to OIDC, to social media, and the Azure Active Directory)
- Authorization (from simple access to RBAC to claims-based authorization)
- Rate limiting
- Response caching
- Test middleware
- Request and response rewriting

Let us look at some simple ASP .NET Core bootstrap code showing middleware providing authentication and authorization:

```
public void Configure(IApplicationBuilder app, IWebHostEnvironment env)
{
  if (env.IsDevelopment())
  {
    app.UseDeveloperExceptionPage(); }
  app.UseRouting();

  app.UseAuthentication();
  app.UseAuthorization();

  app.UseEndpoints(endpoints =>
  {
    endpoints.MapControllers();
  });
}
```

The two highlighted lines inject all the relevant middleware to be able to process authentication and authorization transparently within each request received.

Handling data securely

Many API vulnerabilities relate to the mishandling of data, either in the request or response. At the framework level, consider implementing some of the following best practices:

- Validate input data based on the specified length and type (either in the OAS definition or the request headers)

- If your framework provides type coercion, be wary of implicit assignments (these can lead to mass assignment vulnerabilities)

- Reject any unexpected or invalid content

- Use standard libraries for input validation (such as those provided by OWASP)

- Reject excessively large input data requests or requests for abnormal amounts of data based on normal usage

Error and exceptions

The default response for many frameworks upon encountering an exception is to emit a full stack dump to the output context; for an API, this is likely the response body. While this is helpful for debugging, it is even more helpful for an attacker trying to understand your implementation. Ensure that exception handlers emit the minimum amount of information to your public interfaces and that verbose logs are stored in a secure location. Avoid writing any sensitive information, such as tokens and keys, in your logs.

Return codes

HTTP defines a range of status codes to signal various conditions, from client to server errors. Use specific and descriptive error codes wherever possible to assist your consumers. Generic codes (`200` and `404`, for example) should be avoided except for use for their intended purposes. Judicious use of status codes provides a good compromise between overly verbose error messages and a generic and unhelpful `400 Bad Request` error.

Cross-origin resource sharing (CORS)

If your API will be accessed via a browser using the **XMLHttpRequest** method, then you must implement **cross-origin resource sharing** (**CORS**) in your API. Again, this can be as simple as enabling the appropriate framework feature. For a Micronaut API, this is as simple as the following configuration:

```
---
Micronaut:
  server:
    cors:
      enabled: true
      configurations:
```

```
ui:
  allowed-origins:
    - http://127.0.0.1:8000
```

This example shows another important aspect of CORS; you should always restrict the `allowed-origins` to a specific server (your web application) in preference to using a wildcard (such as `*`), which leaves your API wide open.

Security headers

Finally, your framework should be configured to emit the relevant HTTP security headers to control the behavior of browser-based clients. The recommendations vary quite frequently, and the reader is advised to refer to the excellent OWASP REST security cheat sheet in the *Further reading* section.

Summary

This chapter focused on the critical topic of securing your languages and frameworks when developing APIs. Many of the most serious vulnerabilities we covered in this book originate from vulnerable code implementations or misconfigurations of frameworks. It is vital for the reader to understand the best practices for secure coding and safe framework usage.

First, we examined how to manage the design-first process in practice, covering the important topic of incorporating the OpenAPI definitions in the development lifecycle to ensure that the definition is the source of truth for the API and that all API code is derived from the definition. To do this, we examined two different code-generation tools: Swagger Codegen and the OpenAPI Generator. For the latter, we took a deep dive into understanding how to use it to generate secure server code stubs for various languages. As an API developer, you should become familiar with the fundamentals of generating server implementations for your language of choice, as this will both save you coding time and produce more repeatable and secure code.

Finally, we examined some common patterns for secure configuration and use of frameworks, including topics such as secure defaults, using middleware, handling errors and exceptions, CORS, and security headers. As an API developer, it is important that you understand these basic patterns for securing your frameworks.

In the next chapter, we take the topic of protecting your APIs further as we look at how to use various shield-right methods to further harden and protect your APIs.

Further reading

These links provide further reading on code-generation tools:

* `https://melkornemesis.medium.com/mocks-vs-stubs-choosing-the-right-tool-for-the-job-dbdbc19cf0c5`

- https://openapi-generator.tech/
- https://openapi-generator.tech/docs/generators
- https://openapi-generator.tech/docs/swagger-codegen-migration
- https://spec.openapis.org/oas/v3.1.0#security-scheme-object

These links provide further reading on using OpenAPI with frameworks:

- https://www.npmjs.com/package/swagger-ui-express
- https://ribice.medium.com/serve-swaggerui-within-your-golang-application-5486748a5ed4
- https://learn.microsoft.com/en-us/aspnet/core/tutorials/getting-started-with-swashbuckle?view=aspnetcore-7.0&tabs=visual-studio
- https://learn.microsoft.com/en-us/aspnet/core/tutorials/getting-started-with-nswag?view=aspnetcore-7.0&tabs=visual-studio
- https://github.com/marshmallow-code/apispec/wiki/Ecosystem
- https://blog.ovhcloud.com/openapi-with-python-a-state-of-the-art-and-our-latest-contribution/
- https://springdoc.org/

These links provide further reading on secure API development:

- https://micronaut-projects.github.io/micronaut-security/latest/guide/
- https://cheatsheetseries.owasp.org/cheatsheets/REST_Security_Cheat_Sheet.html
- https://developer.mozilla.org/en-US/docs/Web/HTTP/CORS

11

Shield Right for APIs with Runtime Protection

In the previous chapter, we examined how to secure APIs using best practices for frameworks and languages. While this is important for improving API security, ensuring that your APIs are protected at runtime in production is equally important. This chapter will examine various methods to **shield right** (by shield right, I am referring to various protections for APIs that can be deployed at runtime, as opposed to design or development time) for API security.

First, we will examine basic practices to harden and secure the host platforms your APIs run on, whether Docker containers or operating systems. Then, we will examine the stalwart of runtime defense: the **Web Application Firewall** (**WAF**), and how this can be applied to protect APIs. API gateways and API managers form the core components of your arsenal in protecting APIs, and we will examine in detail the various protections these can bring to your APIs. The final tier of defending APIs is a dedicated API firewall, which protects individual operations, focusing on protecting data according to the OAS definition. With all these components in place, it is important that your APIs are constantly monitored for misbehavior or indicators of attack and that alerts are generated, allowing for a rapid response.

By the end of this chapter, you will understand the key building blocks of a shift-right strategy and how to assemble these to provide optimum API runtime protection.

In a nutshell, this chapter is going to cover the following main topics:

- Securing and hardening environments
- Using WAFs
- Using API gateways and API management
- Deploying API firewalls
- API monitoring and alerting

Technical requirements

For this chapter, you will need a development machine capable of doing the following:

- Running Docker locally

- Running VS Code with various marketplace extensions

- Accessing the internet (you will also need a GitHub account to retrieve the examples)

This chapter contains sample deployments for various runtime protections, such as the Kong API gateway and the 42Crunch API firewall. These configurations and associated instructions will be provided in the GitHub repository for the chapter.

The example code and various breaking changes to the instructions can be found in the `Chapter 11` folder in the book's GitHub repository at the following link: `https://github.com/PacktPublishing/Defending-APIs/tree/main/Chapter11`

Securing and hardening environments

An API server can only be as secure as the foundation upon which it is built. To ensure a strong foundation, several basic best practices should be observed to eliminate the most obvious weaknesses. This section provides summaries of the most important of these best practices for both container images and operating systems. The *Further reading* section provides more detailed references.

Container images

Modern cloud-native development practices have been fueled by the adoption of container technologies as the standard means of application distribution. A container allows an application to be packaged together with all its runtime dependencies and a configured minimalist operating system. This portable package can then be distributed to various runtime environments without concern about dependencies or configuration.

While the adoption of containers has greatly reduced friction between development and operations teams, it has created new vectors of attack that concern security teams. Containers encapsulate a lot of complexity and, if not managed properly, this represents an attack surface.

The first recommendation for hardening containers is an obvious one and something that is easily achieved. Use the minimal possible container image that allows your application to function as intended. There are literally thousands of container images providing fine-tuned images, for example, Python 3.9 on a minimal Ubuntu image. Take the time to research the best minimal image for your needs and use that. If there is not one available, then build your own and contribute this back to the community.

The next recommendation is to ensure that your container image is kept up to date and free of any vulnerabilities. Several excellent open source and commercial tools on the market are designed to address this problem. A few examples are **Clair**, **Docker Bench**, and **Sysdig**. Ensure that you use a container image scanner in your build pipeline to detect any vulnerabilities that may creep in via a vulnerable container image.

Many container-related vulnerabilities occur due to poor runtime configuration of the container execution environment. Fortunately, there are several simple recommendations that can eliminate the bulk of this attack surface:

- **Run the container as a non-root user**: This restricts the *blast radius* in the event of an exploit. The application must be configured to run with a non-privileged user to meet this requirement.

- **Use namespace isolation**: Most container runtimes provide namespace isolation capabilities that can help prevent cross-container access.

- **Use a read-only filesystem**: If your API does not require data to be stored locally (for example, you are using a backend database), then the container filesystem can be mounted in read-only mode. This will prevent any attacker attempting to write data to the filesystem.

- **Disable SETUID and SETGID**: These flags allow the container to elevate its privileges at runtime and can lead to the same security concerns as running as a root user. Disable these flags in the container runtime configuration.

- **Limit resources**: OWASP has identified **API4:2023 – Unrestricted Resource Consumption** as a top API threat, and containers, in particular, can be prone to resource constraints. Fortunately, container runtimes make provision to restrict the amount of available CPU and memory available to container instances. This has the effect of *soft-limiting* the container rather than crashing the operating system.

- **Disable unused ports**: Only expose the ports necessary for your application; leave the others internal to the container only.

Some miscellaneous container assembly tips improve the robustness of container-based solutions. The most important of these tips is never to store credentials or secrets in a container image. Although it may appear that these are hidden because they cannot be easily seen, it does not mean they are safe from prying eyes – simply opening a terminal in a running container instance allows an attacker to read them off the filesystem. Use multi-stage builds to reduce final image size, and use proper versioning of container images (do not use the **:latest** tag).

The *Further reading* section contains excellent resources on how to build high-quality, hardened container images.

Operating systems

The first step toward a hardened *bare-metal* operating system (such as Ubuntu, Fedora, CentOS, Red Hat, and so on) is to choose the most minimal possible installation image necessary for your purposes. For an API server, this would usually involve a server image without any desktop or graphical interfaces. Similarly, after installation, any unused or superfluous services (such as print managers) should be disabled. Also, consider using a distribution with specialized security features such as **Security-Enhanced Linux** (**SE Linux**) or **AppArmor**.

The server should be constantly monitored for suspicious activity, resource, or performance bottlenecks. The best practice is to emit logs to a central location, such as a corporate **Security Incident and Event Manager** (**SIEM**) where specialist teams can monitor them. The server should also be regularly updated in accordance with the agreed update schedule. This typically ensures that all critical patches are applied as soon as they become available.

The next step is to restrict access to the server. Firstly, consider enabling a firewall to prevent unwanted access to ports and services that may be enabled on the server. Create the appropriate access rules to your API from allowed network zones with a secured transport layer enabled. WAFs will be discussed in the next section; if you opt to use one, the built-in operating system WAF may be the most suitable option. Restrict access to the server by enforcing a secure password policy, which includes password complexity and mandatory change requirements. For human users, consider enabling multi-factor authentication. Remove or disable any unused service accounts, particularly those with higher privileges.

Ensure that your API server is run within a *sandboxed* environment that restricts the API server's access to host resources. For instance, ensure that the API server cannot access arbitrary locations on the disk and is restricted from making network connections to unauthorized targets.

Depending on the architecture of your API and data storage, ensure that the server is regularly backed up to prevent data loss.

If you are hosting your API server operating system on cloud infrastructure, selecting a hardened machine image ensures that many of these steps have been taken for you by your cloud provider. For example, hardened images are provided by the **Center for Internet Security** (**CIS**) for most cloud providers. Cloud providers also offer standard, low-friction methods for restricting network or user access, backing up data, and monitoring and logging. If you use cloud hosting, ensure you avail yourself of these features to provide the most secure runtime environment possible.

Using WAFs

A WAF is a **layer 7** device (in the OSI 7 layer model), meaning it operates at the highest layer, namely the application layer. This means that a WAF can interpret HTTP traffic, analyze the payload for threats, and block the traffic accordingly. Physically, a WAF operates like a **reverse proxy**, being located immediately in front of the terminating device (which it is protecting) and receiving all traffic destined for said terminating device. The WAF processes only incoming requests and does not process the response from the server. *Figure 11.1* shows a simplified deployment diagram of a WAF with a rule set filtering API requests to a server.

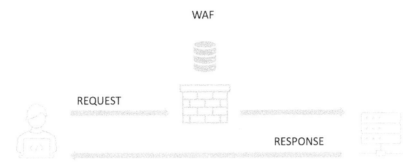

Figure 11.1 – WAF topology

A WAF is configured with a **rule set** that defines the specific rules and policies to be applied. Such rule sets can be defined by the WAF vendor, the community (for example, OWASP defines a rule set for the **ModSecurity** WAF engine), or by the security team in the organization.

In *Chapter 8, Shift-Left for API Security*, we looked at the basics of a WAF in the *Leveraging the positive security model* section. A WAF typically operates in **negative security model** mode, which attempts to block all known bad inputs based on the rule set. This results in a somewhat poor security outcome; namely, the WAF can simultaneously miss valid attacks (false negatives) and trigger on valid inputs (false positives). This leads to a conundrum for security teams operating WAFs, who must choose whether to operate the WAF in **detection mode** only (where it sends alerts on a rule triggering but does not block the traffic) or **protection mode** (where it alerts on a rule triggering and blocks the traffic). This can lead to frustration for application and security teams alike, who are only sometimes confident in the accuracy and protection of the WAF.

WAFs are omnipresent in modern application architectures; for example, here are some typical examples of where a WAF may be implemented:

- At the network edge or perimeter (for example, on the **Cloudflare** network services)

- At the perimeter of cloud deployments (Azure WAF and AWS WAF)

- Host-based deployments where the WAF can be deployed together with the application (for example, a dedicated **Nginx** instance running **ModSecurity** installed as a reverse proxy)

- In a Kubernetes environment as an ingress controller (for example, the **Gloo Edge API Gateway** built on top of **Envoy** provides WAF capability alongside several other security features)

WAFs are the workhorse of the web security industry – if they are available, do deploy and configure them, but be aware of their limitations.

Understanding the Next-Generation Firewall (NGWAF) and Web Application API Protection (WAAP) products

- To address some of the shortcomings of traditional WAFs (principally designed to address the OWASP Top 10 web application vulnerabilities), vendors introduced two new categories of products called **Next-Generation Firewall (NGWAF)** and **Web Application API Protection (WAAP)** products, respectively. We will briefly cover these two products, mainly highlighting how they differ from traditional WAFs.

- NGWAFs attempt to improve on their predecessors by providing more advanced traffic inspection by operating at lower layers of the OSI model. This allows the NGWAF to detect threats based on payloads and signature detection and provides additional capabilities such as **Intrusion Protection Systems (IPSs)**, and application control. Other features include TLS/SSL termination and traffic inspection, bandwidth management, and antivirus inspection.

WAAP is an attempt by vendors to provide more specific protections for applications and APIs in their legacy WAF product. The specific areas of focus include the following:

- **Malicious bot protection**: Identifies bot attacks and blocks them in real time

- **Distributed denial-of-service (DDoS) attacks**: Uses sophisticated traffic analysis to identify distributed attacks targeting APIs and applications

- **Advanced rate limiting**: Implements rate limiting to protect APIs and applications

- **Protection for APIs**: Provides additional context around an API to provide specific API protections for vulnerabilities such as broken-object-level authorization and broken-function-level authorization

- **Runtime application self-protection**: Integrates a protection layer directly into the application at the instrumentation layer to protect the API and application

- **Account takeover protection**: Protects against many common account takeover attacks using password lists or leaked credentials

This is an aspirational set of capabilities, and WAAP technology certainly promises a wealth of advanced protection. Unfortunately, at the time of writing, this technology has failed to deliver on its initial promise, with solutions often plagued by difficulties in deployment and challenges in tuning against real-world data. Many vendors offering WAAP make it difficult to evaluate the technology by not providing a community or trial version. As such, we will limit the discussion on WAAP to this basic summary. You are advised to keep an eye on this product category – it certainly promises a lot.

Using API gateways and API management

We now focus on the core *shield right* technology for APIs, namely API gateways and **API management (APIM)** solutions. To understand what an API gateway does, let us consider how APIs were deployed before gateways existed. Typically, an API would be instantiated on a server, assigned a resolvable name, and connected to the public internet. While this achieved the result of bringing the API online, it created a myriad of other problems for system administrators:

- Difficulty in scaling the service, either horizontally or vertically

- A very tightly coupled architecture – the internal architecture of the system was exposed directly to the client and could not be refactored without potentially breaking all clients

- The lack of a common approach to **cross-cutting concerns** (issues common to all APIs best addressed in a standard method) meant that each API had to implement its own logging, access control, rate limiting, and load balancing

API gateways solve this problem by acting as a gateway between external clients on a public interface and the internal implementation (typically a collection of APIs or microservices). The gateway processes requests from the public interface and, based on the request, determines how to route this internally. A typical API gateway solution is shown in *Figure 11.2*.

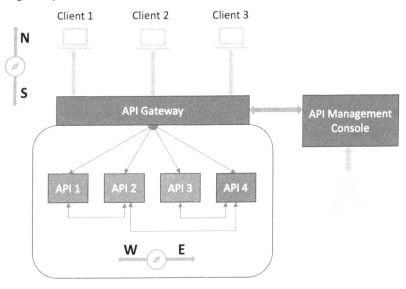

Figure 11.2 – API gateway/management deployment

In this example, external traffic from the three clients is routed via the gateway to a composite of internal APIs – the so-called **north-south** traffic route (external to internal traffic). A series of internal APIs provide the backend services on the **south** end of the API gateway. These APIs may also communicate internally with one another as part of the overall microservice design. The communication is internal only, referred to as **west-east** traffic, and does not pass through the API gateway.

To understand the value that an API gateway brings from a security point of view, let us examine some of the common features of a typical API gateway:

- **Routing policy**: Functionally, the most important feature of an API gateway is the ability to apply dynamic routing policies that decouple clients from the internal implementation. For instance, an API endpoint may stay the same externally, but internally, this can now refer to a new version of the API or a composite of services.

- **Protocol translation**: Another important function a gateway provides is translating from one protocol to another. A common example is an internal legacy system running the **Simple Object Access Protocol** (**SOAP**) protocol, which can be modernized by converting to **gRPC Remote Procedure Calls** (**gRPC**). Rather than rewriting the backend, the gateway can perform this translation on the fly.

- **Caching**: As an aggregation point for access to APIs, the gateway is ideal for performing a central caching function, providing caching management to all subscribers, and offloading this complex function from the individual APIs.

- **Load balancing**: As well as managing traffic caching, the API gateway can also offer central load balancing services to ensure that all incoming requests are distributed fairly across resources based on availability.

- **Authentication and authorization**: API gateways are perfectly equipped to perform authentication and authorization at a central enforcement point rather than at the individual API level. Gateways can examine API keys and tokens in the header or body of requests and grant access to downstream APIs based on a central policy. Gateways also offer advanced features such as JWT token validation and token exchange to allow access to internal APIs.

- **Security features**: API gateways can provide many dedicated security features such as **Cross-Origin Resource Sharing** (**CORS**) enforcement to protect against leakage of information from APIs, bot detection and protection, JWT inspection and validation, data loss protection (by inspecting for **Personally Identifiable Information** (**PII**) at the gateway level), and rate-limiting functions.

- **Logging and aggregation**: A central logging capability is another security-related function suited to an API gateway. Gateways can log and aggregate all requests and responses in a central log management facility. This is a great example of how an API gateway solves a cross-cutting concern, providing a transparent and uniform logging function to all APIs without implementing their own logging function.

- **Analytics and monitoring**: Similar to logging, gateways can provide inspection of API traffic patterns and present these in a centralized manner, typically via dashboards or analytics reporting engines.

- **Rate limiting**: Since gateways terminate all external API traffic and manage sessions, they are the ideal enforcement point for rate limiting and protecting their backend APIs from excessive requests. Gateways are designed to be robust, capable of handling high traffic volumes, and a natural place to enforce rate limits.

- **Access control**: Finally, as the boundary to the outside world, an API gateway is the perfect place to implement access control to the internal APIs. Access restrictions are typically based on IP addresses (individual IP addresses or ranges of high-risk IP addresses such as TOR exit nodes) and client fingerprints.

- Now that we have an overview of the types of capabilities offered by an API gateway let us explore in great detail some of the security features of the community version of the popular Kong API gateway.

API gateway versus APIM

The terms API gateway and API management are used somewhat interchangeably. The simplest way to think about them is that the API gateway is the data plane handling API traffic, and API management functions more as the control plane managing policies, versions, releases, and so on.

The API gateway terminates all requests from the outside network, processes them, including validating security constraints such as authentication and authorization, forwards them to the internal services, and processes the responses to the client. The API gateway sits directly in line with API requests and responses.

The API management function provides – as the name suggests – management capabilities over APIs, including controlling the versions of APIs, monitoring API usage and consumption metrics, controlling consumption quotas, controlling API gateway policies, and developer or consumer features such as an API catalog.

Figure 11.2 shows this relationship quite clearly – the API gateway sitting in line with traffic, and the API management console controlling the configuration.

Implementing security patterns in the Kong API gateway

One of the more popular API gateways on the market is **Kong Gateway** (`https://konghq.com/install#kong-community`), available in a community (open source) version that provides many core features without enterprise management capabilities. For this section, the examples will show how the core security features can be used to protect APIs.

First, let us understand the architecture of Kong Gateway. *Figure 11.3* shows the external client, the **routes** and **services**, and the backend servers.

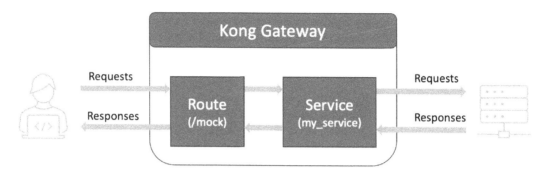

Figure 11.3 – Kong Gateway architecture

The route defines how the gateway responds to incoming requests and routes them accordingly. In my example, I have defined three routes: /defendingapis, /mock, and /pixi, and my instance is running on http://localhost:8000/. *Figure 11.4* shows this configuration in the management console in Kong.

Routes

A Route defines rules to match client requests, and is associated with a Service. **Learn more** ⤤

Name ⌄	Protocols	Hosts	Methods	Paths	Tags	
defendingapis_route	http https	-	-	/defendingapis	-	⋮
mockbin_route	http https	-	-	/mock	-	⋮
pixi_route	http	-	-	/pixi	-	⋮

Figure 11.4 – Kong Gateway route definitions

In this configuration, the gateway will respond to requests starting with those path prefixes and ignore anything else. For example, this is the error if a client accesses a non-existent path:

```
colind@mbm1pro: ~/API Demos # curl http://localhost:8000/my_bad_path
{
    "message":"no Route matched with those values"
}
```

The gateway services define the upstream service to be accessed by Kong Gateway. *Figure 11.5* shows how these upstream services are defined. In my example, I have three services, one for each route (this is a simplification; Kong allows a complex composition of services per route based on path, protocol, etc.).

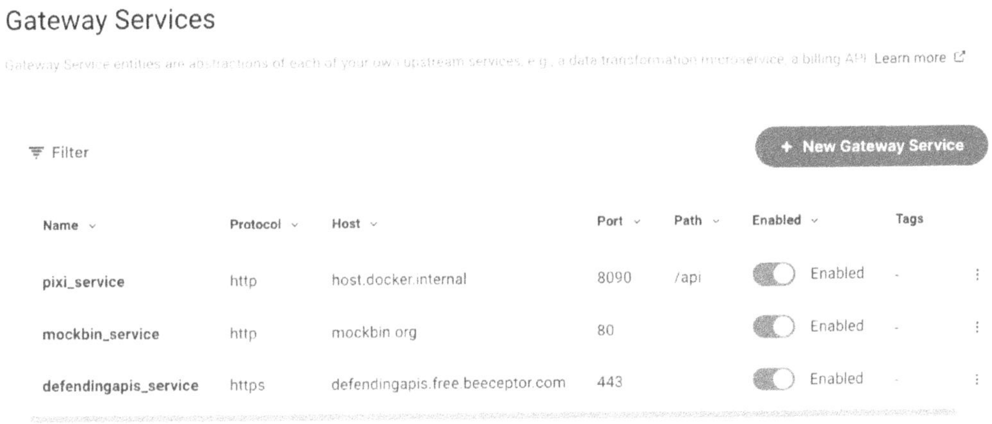

Gateway Services

Gateway Service entities are abstractions of each of your own upstream services, e.g., a data transformation microservice, a billing API. Learn more

Name	Protocol	Host	Port	Path	Enabled	Tags	
pixi_service	http	host.docker.internal	8090	/api	◉ Enabled	-	⋮
mockbin_service	http	mockbin.org	80		◉ Enabled	-	⋮
defendingapis_service	https	defendingapis.free.beeceptor.com	443		◉ Enabled	-	⋮

Figure 11.5 – Kong Gateway service definitions

To understand how the gateway definition works, consider the final entry for the `defendingapis_service`. This will map `https` access on port `443` to the upstream API service at the address `defendingapis.free.beeceptor.com`. (There is a very useful API mocking service hosted at this link: `https://beeceptor.com/console/defendingapis`. I have defined a few endpoints with various payloads for demonstration purposes.)

To see how the gateway and route work together, I make the following request to the Kong Gateway local instance:

```
colind@mbm1pro: ~/API Demos # curl http://localhost:8000/
defendingapis/info
{
    "status": "success",
    "user": {
        "name": "Colin Domoney",
        "email": colin.domoney@gmail.com,
        "active": true
        }
}
```

Kong interprets the `defendingapis` base path as the route and maps this to `defendingapis_service`, which then makes an outbound connection to `defendingapis.free.beeceptor.com`, prepending the `/info` path to the request. Putting this all together results in a request to `https://defendingapis.free.beeceptor.com/info`, which is a dummy endpoint hardcoded to return the JSON response shown. Feel free to try this URL and confirm for yourself.

This shows how the gateway can intercept request and response traffic and create URL translations based on the route and services. This is a simple one-to-one example, but it is possible to create complex mappings based on paths, protocols, and ports.

Out of the box, Kong does not provide much other functionality, but where it comes into its own is with the power of plugins, which extend the basic capabilities. Kong offers its plugin marketplace (`https://docs.konghq.com/hub/`), providing capabilities for authentication, security, traffic control, serverless, analytics and monitoring, transformations, and logging.

Plugins can be installed globally, within a route, or within a service. *Figure 11.6* shows the four example plugins we will use in this section installed globally.

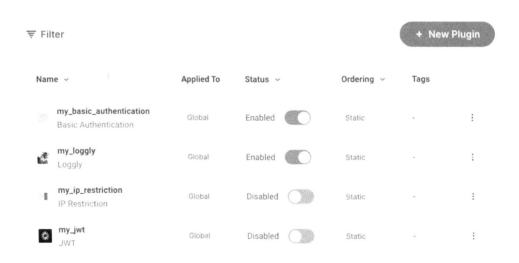

Figure 11.6 – Kong Gateway plugins

Now that we understand the basics of the Kong Gateway architecture, let us examine how the various plugins can provide security features.

Basic authentication

To enforce either basic authentication (or JWT validation), the user credentials must be stored on Kong in the **Consumers** section, as shown in *Figure 11.7* where both basic authentication and JWT credentials are defined.

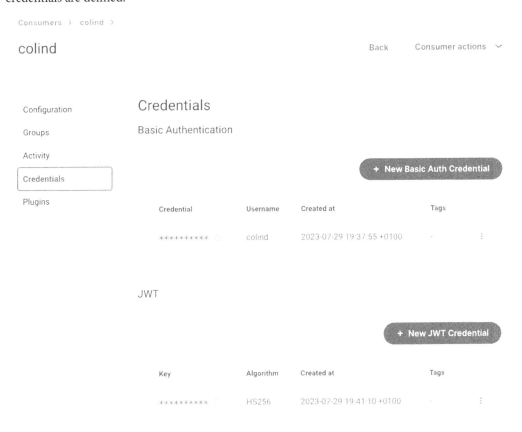

Figure 11.7 – Kong consumer credentials

To enforce basic authentication checks, enable the plugin and attempt to access the API as shown in the following code:

```
colind@mbm1pro: ~/API Demos # curl http://localhost:8000/
defendingapis/info
{"message":"Unauthorized"}

colind@mbm1pro: ~/API Demos # curl http://localhost:8000/
defendingapis/info -H "Authorization: Basic Y29saW5kOmFwaXNhcVmdW41"
{"message":"Invalid authentication credentials"}

colind@mbm1pro: ~/API Demos # curl http://localhost:8000/
defendingapis/info -H "Authorization: Basic Y29saW5kOmFwaXNhcVmdW4="

{
    "status": "success",
    "user": {
        "name": "Colin Domoney",
        "email": "colin.domoney@gmail.com",
        "active": true
    }
}
```

The first call provides no credentials and returns an `"Unauthorized"` error message. The second call provides incorrect credentials, and `"Invalid authentication credentials"` is returned. The third call provides the correct credentials (in Base64, which you can decode if you wish), and access to the API is granted. Note that the API itself (hosted at `https://defendingapis.free.beeceptor.com/info`) does not provide the authentication check – this is done entirely in the Kong Gateway.

JWT validation

To enforce JWT validation, two items need to be configured on the Kong gateway. Firstly, the algorithm and signing key must be stored in the consumer credentials, as shown in *Figure 11.8*. The algorithm can either be symmetric (as is the case for HS256 shown) or asymmetric, in which case the private key must be entered.

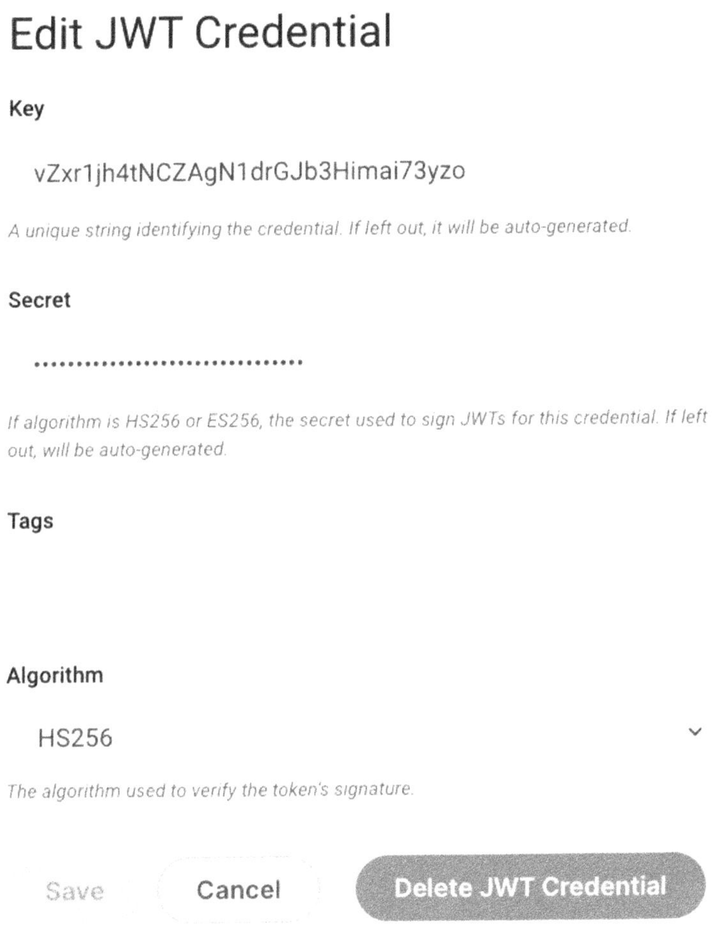

Edit JWT Credential

Key

vZxr1jh4tNCZAgN1drGJb3Himai73yzo

A unique string identifying the credential. If left out, it will be auto-generated.

Secret

....

If algorithm is HS256 or ES256, the secret used to sign JWTs for this credential. If left out, will be auto-generated.

Tags

Algorithm

HS256   ⌄

The algorithm used to verify the token's signature.

Save   Cancel   **Delete JWT Credential**

Figure 11.8 – JWT credentials on a Kong Gateway

In the JWT plugin, the following items need to be specified, as shown In *Figure 11.9*:

- The claims to be verified – in this case, the expiration in the exp field
- The location where the token is transmitted – in this case, a header called authorization
- The field specifying the signing key – in this case, an iss field

Figure 11.9 – JWT validation configuration

To enforce basic authentication checks, enable the plugin and attempt to access the API as shown in the following code segment:

```
colind@mbm1pro: ~/API Demos # curl http://localhost:8000/
defendingapis/info
{"message":"Unauthorized"}

colind@mbm1pro: ~/API Demos # curl http://localhost:8000/
defendingapis/info -H "Authorization: Bearer eyJ0eXAiOiJKV1QiLCJhbG
ciOiJIUzI1NiJ9.eyJyb2xlIjoiQWRtaW4iLCJpc3MiOiJ2WnhyMWpoNHROQ1pBZ04xZHJ
HSmIzSGltYWk3M3l6byIsImV4cCI6MTY5MzMzMTQwMSwiaWF0IjoxNjkwNjU2NjAxLCJ1
c2VybmFtZSI6ImNvbGluZCJ9.-k3ImcHtw22G25ooVOnNiulz65hjfOyo0ou5wspR911"
{"message":"Invalid signature"}

colind@mbm1pro: ~/API Demos # curl http://localhost:8000/
defendingapis/
info -H "Authorization: Bearer eyJ0eXAiOiJKV1QiLCJhbGGciOiJIUzI1NiJ9.
eyJyb2xlIjoiQWRtaW4iLCJpc3MiOiJ2WnhyMWpoNHROQ1pBZ04xZHJHSmIzSGltYWk3M3
l6byIsImV4cCI6MTY5MzMzMTQwMSwiaWF0IjoxNjkwNjU2NjAxLCJ1c2VybmFtZSI6ImNv
bGluZCJ9.-k3ImcHtw22G25ooVOnNiulz65hjfOyo0ou5wspR9oo"
{
    "status": "success",
     "user": {
         "name": "Colin Domoney",
         "email": colin.domoney@gmail.com,
         "active": true
        }
}
```

The first call provides no JWT token and returns a "Unauthorized" error message. The second call provides an invalid JWT token, and "Invalid signature" is returned. The third call provides a correct JWT token (which you can decode if you wish), and access to the API is granted. Note that the API itself (hosted at `https://defendingapis.free.beeceptor.com/info`) does not provide the authentication check – this is done entirely in the Kong gateway.

IP address restriction

IP address restriction is one of the simplest access controls on the Kong gateway. This is configured by providing a list of allowed IP addresses, followed by a list of denied IP addresses, as shown in *Figure 11.10*.

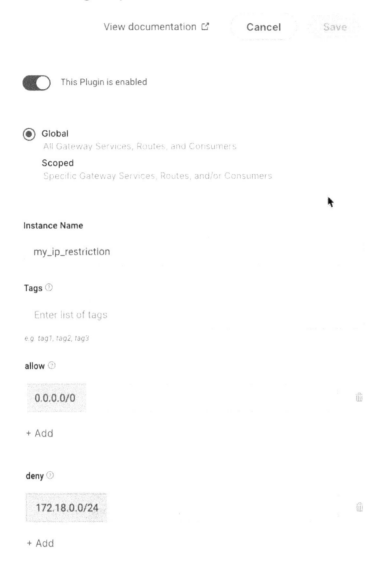

Figure 11.10 – IP address restriction

In this example, all IP addresses are allowed via the $0.0.0.0/0$ CIDR range, and then the $172.18.0.0/24$ CIDR range is explicitly blocked (this is the local Docker network range hosting my test environment – effectively, I am blocking access to my target APIs).

Accessing the upstream API with the IP address restriction disabled and then enabled is shown here:

```
colind@mbm1pro: ~/API Demos # curl http://localhost:8000/
defendingapis/info
{
    "status": "success",
     "user": {
         "name": "Colin Domoney",
         "email": colin.domoney@gmail.com,
         "active": true
        }
}

colind@mbm1pro: ~/API Demos # curl http://localhost:8000/
defendingapis/info
{
  "message":"Your IP address is not allowed"
}
```

This is a basic access control implemented purely at the gateway level. Similar plugins exist to perform rate limiting.

Logging

Finally, let us examine how easy it is to connect the Kong gateway to a central logging function. In this instance, I will use the **Loggly** logging platform.

The plugin is configured with some basic parameters, such as the Loggly host, API token, and the desired logging level (in this case, we log at the info level to capture everything). This is shown in *Figure 11.11*.

Figure 11.11 – Loggly plugin configuration

Once configured, every transaction on the Kong gateway matching the criteria will be logged to the Loggly platform in **syslog** format. A sample entry is shown in *Figure 11.12*.

□ 2023-08-05 16:22:53.000
                % Copy link   ↗ Expand Event   ↙ Create Derived Fields   ⊛ View Surrounding Events   ⋮

  tag: kong   logtype: json   logtype: syslog
□ json:
    client_ip: 172.18.0.1
    started_at: 1691248973686
    upstream_status: 400
    upstream_uri: /api/user/register
    workspace: 161599c1-4999-4407-bcf1-4e23345b914a
  ⊕ latencies: { proxy: 36, request: 47, kong: 11 }
  ⊕ request: { headers: { content-length: "128", host: "host.mbm1pro.home:8000", content-type:
      "application/json", user-agent: "curl/7.88.1", accept: "application/json" }, method: "POST", size:
      299, uri: "/pixi/user/register", url: "http://host.mbm1pro.home:8000/pixi/user/register" }
  ⊕ response: { headers: { date: "Sat, 05 Aug 2023 15:22:53 GMT", content-length: "40", x-kong-upstream-
      latency: "36", x-powered-by: "Express", x-kong-proxy-latency: "11", etag:
      "W/"28-/YFCZ9Puaf3vQiVNC74ZLZGAmK4"", content-type: "application/json; charset=utf-8", connection:
      "close", via: "kong/3.3.1.0-enterprise-edition" }, size: 354, status: 400 }
  ⊕ route: { request_buffering: true, path_handling: "v0", https_redirect_status_code: 426, created_at:
      1691246101, preserve_host: false, strip_path: true, response_buffering: true, updated_at:
      1691247994, paths: ["/pixi"], service: { id: "3edba969-d76f-47ea-a655-aae52467bd26" }, name:
      "pixi_route", regex_priority: 0, id: "f3763a2e-8527-480b-8eb6-f58dad105bc2", protocols: ["http"],
      ws_id: "161599c1-4999-4407-bcf1-4e23345b914a" }
  ⊕ service: { connect_timeout: 60000, created_at: 1691245991, enabled: true, retries: 5, path: "/api",
      protocol: "http", updated_at: 1691248903, port: 8090, write_timeout: 60000, name: "pixi_service",
      host: "host.docker.internal", id: "3edba969-d76f-47ea-a655-aae52467bd26", read_timeout: 60000,
      ws_id: "161599c1-4999-4407-bcf1-4e23345b914a" }
  □ tries:
    ⊕ [0]: { balancer_latency_ns: 56832, port: 8090, balancer_start: 1691248973697, ip:
        "192.168.65.254", balancer_latency: 0, balancer_start_ns: 1691248973697100000 }
□ syslog:
    appName: kong
    facility: user-level messages
    host: 2709b4c9e350
    priority: 14
    severity: Informational
    timestamp: 2023-08-05T15:22:53Z

Figure 11.12 – Sample Loggly log entry

This record shows the level of transaction detail in both the request and response, making it possible for the support team to analyze any API transactions when investigating an incident.

This centralized logging on a dedicated platform such as Loggly shows the value of logging at the API gateway instead of in individual APIs. Firstly, the API developers do not have to add any code to their APIs; instead, the gateway handles this transparently as a cross-cutting concern. Secondly, the logging can be easily centralized to a central platform, eliminating the need to manage many distributed logging platforms.

Finally, with all transactions being processed at the API gateway, it is easy to get an oversight of API operations at a composite level in the organization. *Figure 11.13* shows a summary view of passing and failing operations over time for all routes – from here, it is very easy to spot trends or patterns requiring further attention.

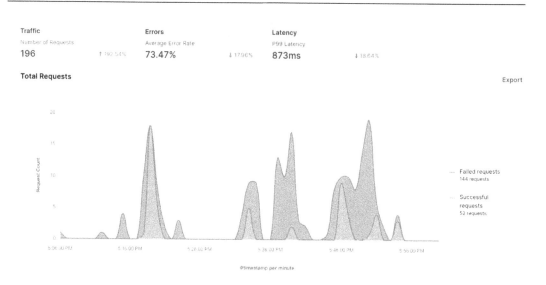

Figure 11.13 – Kong API gateway traffic analysis

This section shows the variety of protections that even an open source version of an API gateway can offer APIs – if you have an API gateway on your platform, make sure it is activated and configured appropriately for optimal API security.

Best practices for API gateway protection

To conclude this chapter on API gateways, let us look at some specific recommendations from Microsoft to address the OWASP API Security Top 10 2019. Although the advice targets their Azure APIM specifically, many key recommendations apply to any gateway offering the relevant features. Their online guide (`https://learn.microsoft.com/en-us/azure/api-management/mitigate-owasp-api-threats`) provides a wealth of further recommendations for the use of API gateways and is worth reading in detail.

API1:2019 – Broken object-level authorization

In *Chapter 9, Defending against Common Vulnerabilities*, the *Authorization vulnerabilities* section covered the best practices for protecting against **Broken Object-Level Authorization** (**BOLA**), and the number one recommendation is to implement explicit authorization against the requested object in the API code. This is also Microsoft's overarching guidance and should be followed as far as possible.

An API gateway can only partially mitigate some of the risks of BOLA using the following technique: use a translation policy to remap identifiers from the request to the backend and vice versa. This prevents the internal identifiers from being exposed externally and vulnerable to enumeration.

This is only a partial remedy and is not sufficient on its own.

API2:2019 – Broken user authentication

An API gateway effectively implements various authentication methods centrally, as covered in the section on the Kong gateway.

For example, the following are supported out of the box in most API gateways:

- Basic authentication
- API keys and tokens
- Client certificates
- JWT tokens

Designers can offload this responsibility from their API backends by choosing to authenticate at the API gateway level. The best practice is to perform a secondary authentication, perhaps using JWT tokens in the **phantom token** or **split token** patterns.

API3:2019 – Excessive data exposure

The recommendation to avoid excessive data exposure is to ensure that the API backend only returns the minimum amount of data to meet the business function of the API and no more.

The API gateway can offer additional protection against excessive data exposure, such as the following:

- Use **revisions** and **versions** when the data format changes to handle deprecations gracefully and avoid breaking clients
- Some gateways can perform payload transformations, which can be used to mask or remove superfluous data
- Perform content validation using API schema such as the OpenAPI definition
- Validate the headers returned in the response and strip any that are unnecessary

API4:2019 – Lack of resources and rate limiting

The API gateway performs excellently at protecting against a lack of resources or rate limiting. In most instances, the API gateway is the preferred enforcement point for this protection, thereby removing the burden from the API backend.

The following are some examples of how an API gateway can protect against this threat:

- Use short- and long-term limiting policies to restrict the number of API calls per client.
- Ensure that OpenAPI definitions enforce the maximum allowable values for page sizes, maximum lengths, and regular expression lengths for strings. Use the gateway to enforce these values.
- Leverage the caching features of the gateway to avoid hitting the API backend unless necessary.

- Guard against JWTs claiming resources outside their scope and block clients who abuse their tokens or keys.

- Implement a CORS policy to prevent inadvertent leakage of resources to other domains than intended. Avoid the use of wildcard operators in CORS policies.

- Be sure to fully examine all the available resource- and rate-limiting features of your chosen API gateway, and ensure they are enabled to protect your backend APIs.

API5:2019 – Broken function-level authorization

In *Chapter 9, Defending against Common Vulnerabilities*, the *Authorization vulnerabilities* section covered the best practices for protecting against **Broken Function-Level Authorization** (**BFLA**), and the number one recommendation is to implement explicit authorization against the requested function or operation in the API code. This is also Microsoft's overarching guidance and should be followed as far as possible.

The only useful recommendation at the API gateway level is to avoid using wildcard paths in constructing the gateway's route and service table. Often, routes might be constructed with a catch-all wildcard path, which opens the possibility of exposing all endpoints in the backend API. Rather, use explicit mapping (such as /user, /orders, etc.) to map the allowed endpoints, making enumeration and discovery harder for an attacker.

API6:2019 – Mass assignment

The recommendation to avoid excessive data exposure is to ensure that the API backend only returns the minimum amount of data to meet the business function of the API and no more.

The API gateway can offer additional protection against excessive data exposure, such as the following:

- Some gateways can perform payload transformations that can be used to mask or remove superfluous data

- Perform content validation using API schemas such as the OpenAPI definition

API9:2019 – Improper assets management

The API gateway and its management element (typically referred to as an APIM) offer administrators various facilities to restrict and control the APIs deployed publicly. A typical organizational pattern is to ensure that API teams can only deploy APIs if they are provisioned through an APIM/gateway, and are accompanied by an OpenAPI definition. This allows administrators to control which versions of which APIs are deployed live and can help identify shadow and zombie APIs. Of course, this only works if sufficient governance is placed around this process.

API10:2019 – Insufficient logging and monitoring

As discussed in the section on logging in the Kong gateway, the API gateway is the optimal place to implement centralized and uniform logging.

There are no specific recommendations for either *API7:2019 – Security Misconfiguration* or *API8:2019 – Injection* at the API gateway level.

Deploying API firewalls

While the API gateway is the workhorse of API runtime protection, there are some specific drawbacks to being solely reliant on the API gateway for protection:

- Not all the OWASP API Security Top 10 vulnerabilities can be protected at the gateway (for example, injection and security misconfiguration), and for data-based vulnerabilities, protection is only partial.

- Due to its centralized function in line with all API traffic, the API gateway can be a performance bottleneck, particularly when doing deep packet inspection, filtering, and transformation.

- API gateways typically protect North-South traffic (refer to *Figure 11.2*) and do not protect West-East traffic, namely intra-API traffic between microservices.

- While the gateway is ideal for enforcing central policies, there may be occasions when API developers wish to deploy local policies alongside their API, for example, validating JWT tokens for excessive claims or performing local rate-limiting.

Fortunately, there is an ideal solution to address these concerns – the API firewall. There are a few commercial vendors providing such products, such as **Wallarm** and **42Crunch**. In this section, we will take a quick look at the 42Crunch firewall since this is available as a limited trial by registering on their website.

Architecturally speaking, the firewall acts as a reverse proxy in front of the API backend. The firewall terminates the inbound connection, processes it against a ruleset based on the OpenAPI definition for the API, passes it to the API, and processes the response through the same ruleset before returning it to the client. Conceptually, it acts in very much the same way an API proxy does.

The main difference is that the API firewall enforces the OpenAPI definition of the API and provides some other security features, such as removing extraneous headers, validating JWT tokens, and performing local rate limiting.

The 42Crunch firewall is normally deployed simultaneously as the API, often in a **sidecar** pattern in Kubernetes. The 42Crunch PaaS platform configures the firewall (log levels, blocking mode, and OpenAPI definition) and can then be totally disconnected if desired. The firewall can log transactions to the 42Crunch platform or to a local filesystem. The simplified architecture is shown in *Figure 11.14*.

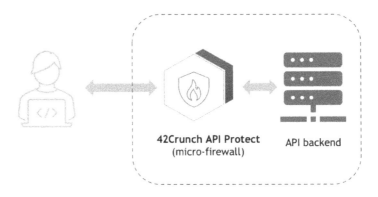

Figure 11.14 – 42Crunch API firewall

To see how the firewall protects an API, let us consider an endpoint called /user/edit_info within an API; the OpenAPI definition for this endpoint is shown in *Figure 11.15*:

```
"/user/edit_info": {
  "put": {
    Scan | Try it | Audit
    "tags": [
      "users"
    ],
    "summary": "edit user information",
    "description": "user supplies valid token and updates all user info",
    "operationId": "edituserinfo",
    "parameters": [
      {
        "name": "x-access-token",
        "in": "header",
        "required": true,
        "schema": {
          "maxLength": 1000,
          "pattern": "^([a-zA-Z0-9_=]{4,})\\.([a-zA-Z0-9_=]{4,})\\.([a-zA-Z0-9_\\-\\+\\/=]{4,})",
          "type": "string"
        }
      }
    ],
    "requestBody": {
      "description": "userobject",
      "content": {
        "application/json": {
          "schema": {
            "$ref": "#/components/schemas/UserUpdateData"
          }
        }
      },
      "required": true
    },
```

Figure 11.15 – Definition of the /user/edit_info endpoint

Tracing through to the `UserUpdateData` object, we see the definition in *Figure 11.16*.

```
"UserUpdateData": {
  "required": [
    "email",
    "name"
  ],
  "type": "object",
  "additionalProperties": false,
  "properties": {
    "email": {
      "maxLength": 320,
      "minLength": 6,
      "pattern": "^(?:[\\w\\-+!#$%&'*/=?^`|{}~]+(?:\\.[\\w\\-+!#$%&'*/=?^`|
      "type": "string",
      "format": "email",
      "example": "foo@bar.com",
      "x-42c-format": "o:email"
    },
    "name": {
      "maxLength": 30,
      "minLength": 5,
      "pattern": "^[\\w\\s\\.]{5,30}$",
      "type": "string"
    },
    "account_balance": {
      "maximum": 1000,
      "minimum": -50,
      "type": "number",
      "format": "float"
    }
  }
},
```

Figure 11.16 – UserUpdateData definition

The most important takeaway is that this endpoint should only accept requests with a JSON body containing email and name, and nothing else.

Now, let us examine the behavior of the unprotected API with an invalid input body; in this case, we are attempting a mass assignment attack by supplying an extra parameter – is_admin. This is shown in *Figure 11.17*.

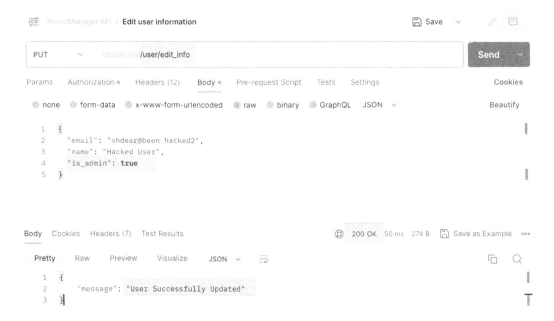

Figure 11.17 – Response from an unprotected API

Despite the invalid input, we have received a 200 OK status code and a message indicating success. This is a bug in the API implementation; this additional input should have been rejected, and an error code returned.

If we repeat this request against a protected instance of the same API, we get quite a different result, as shown in *Figure 11.18*.

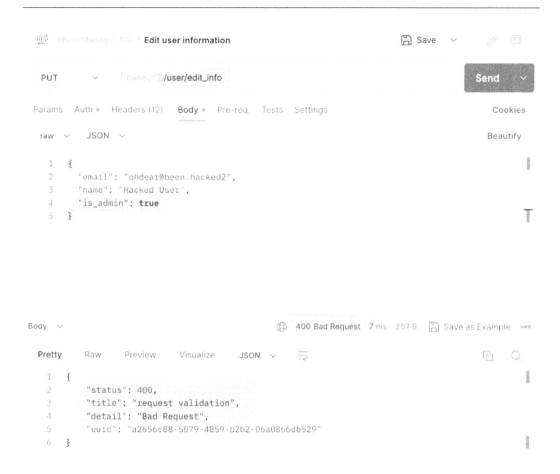

Figure 11.18 – Response from a protected API

While the API backend has not changed, the response is quite different. A 400 Bad Request status code has been returned, together with a response body indicating an error with the request. The firewall intercepted the request and found an additional field (the is_admin field), terminated the request, and returned the error. The request never even reached the API backend.

On the 42Crunch PaaS platform, the error log shows the transaction and provides more details about the fault condition, as shown in *Figure 11.19*.

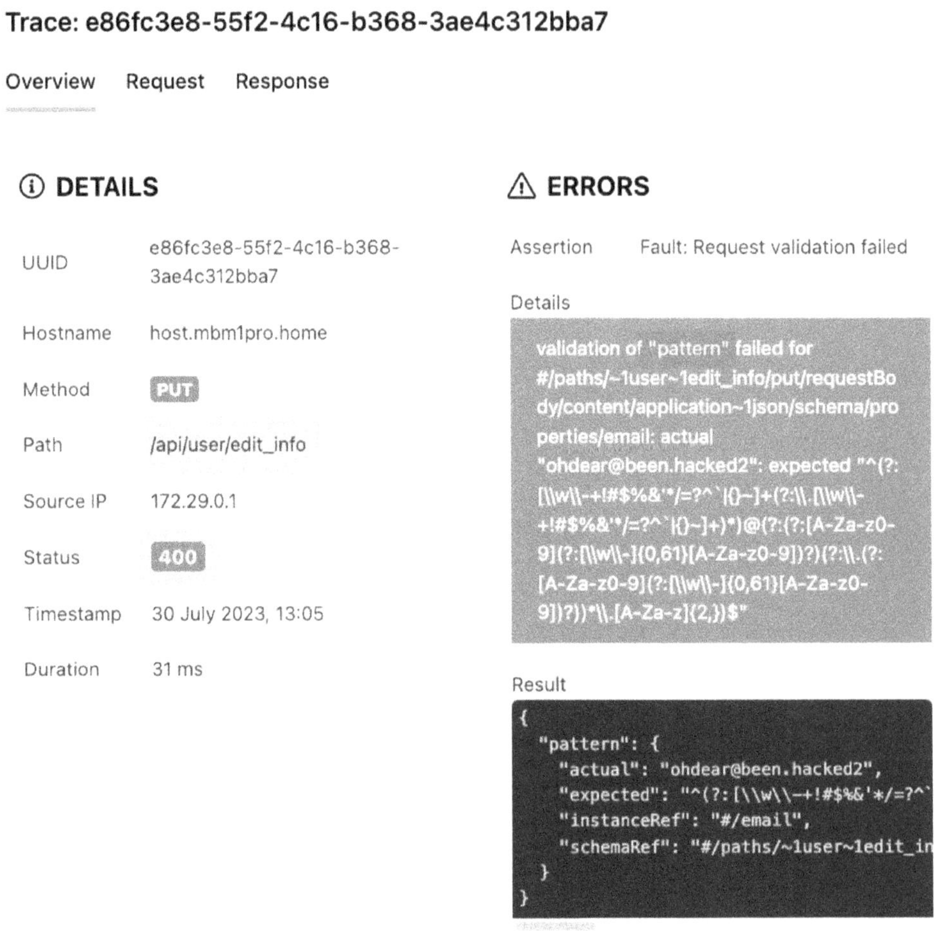

Trace: e86fc3e8-55f2-4c16-b368-3ae4c312bba7

Overview   Request   Response

ⓘ DETAILS

UUID	e86fc3e8-55f2-4c16-b368-3ae4c312bba7
Hostname	host.mbm1pro.home
Method	PUT
Path	/api/user/edit_info
Source IP	172.29.0.1
Status	400
Timestamp	30 July 2023, 13:05
Duration	31 ms

⚠ ERRORS

Assertion   Fault: Request validation failed

Details

validation of "pattern" failed for #/paths/~1user~1edit_info/put/requestBody/content/application~1json/schema/properties/email: actual "ohdear@been.hacked2": expected "^(?:[\\w\\-+!#$%&'*/=?^`|{}~]+(?:\\.[\\w\\-+!#$%&'*/=?^`|{}~]+)*)@(?:(?:[A-Za-z0-9](?:[\\w\\-]{0,61}[A-Za-z0-9])?)(?:\\.(?:[A-Za-z0-9](?:[\\w\\-]{0,61}[A-Za-z0-9])?))*\\.[A-Za-z]{2,})$"

Result

```
{
  "pattern": {
    "actual": "ohdear@been.hacked2",
    "expected": "^(?:[\\w\\-+!#$%&'*/=?^`
    "instanceRef": "#/email",
    "schemaRef": "#/paths/~1user~1edit_in
  }
}
```

Figure 11.19 – 42Crunch API firewall log entry

This shows the power of a dedicated API firewall in providing precise enforcement of the OpenAPI contract in direct proximity to the API backend.

API monitoring and alerting

In this final brief section, we will look at how to monitor an API within a SIEM and **Security Operation Center (SOC)** using as an example the 42Crunch API firewall and the Microsoft **Sentinel** SIEM.

The 42Crunch firewall emits logs to a local filesystem that can be collected by a log forwarder and forwarded to **Azure Log Analytics** for ingestion into Microsoft Sentinel. This simplified architecture is shown in *Figure 11.20*.

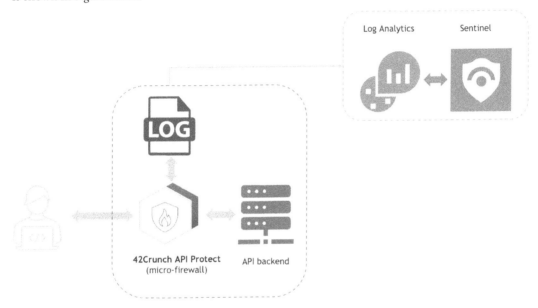

Figure 11.20 – 42Crunch firewall log ingestion in Sentinel

Using the 42Crunch marketplace extension, Sentinel can process the API logs and alert against 12 active API rules, as shown in *Figure 11.21*.

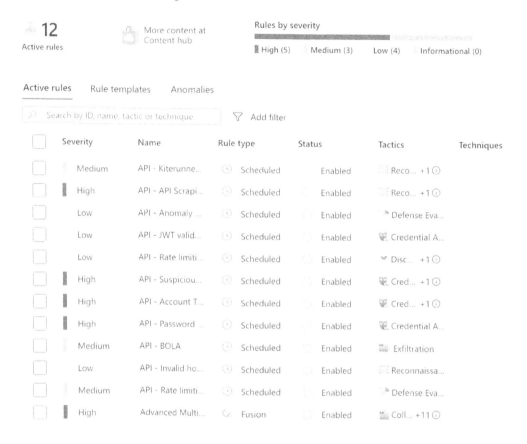

Figure 11.21 – Sample Sentinel API firewall rules

When a rule is triggered, this is recorded on Sentinel as an incident and annotated with all the instance data, such as source IP address, destination path and port, response and request bodies, and return status code. An example of ongoing incidents is shown in *Figure 11.22*.

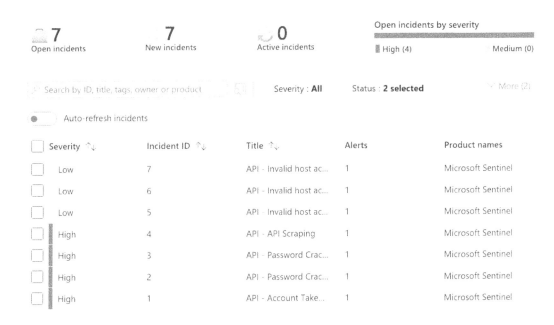

Figure 11.22 – Sentinel API incidents

Finally, the API log data can be presented in various customizable workbooks as shown in *Figure 11.23*.

Figure 11.23 – 42Crunch API workbook sample

The advantages of collating all API security logs in a central industry standard SIEM such as Sentinel include the following:

- **Cost reduction**: Avoiding duplication of costs associated with buying and operating a dedicated API security monitoring tool, and instead add to the value of your existing investment in SIEM/ SOC solutions by enriching them with API logs and alerts.

- **Accuracy**: Using a dedicated API micro-firewall capable of inspecting API traffic at the API level (layer 7) against an OAS definition rather than relying on network traffic inspection (layer 4).

- **Simplicity**: The biggest cost with security operations is the SOC operators and analysts. Surfacing API logs and alerts into existing SOCs avoids the complexity of operating a separate platform.

- **Integration**: For Azure users, direct integration with, for example, firewalls, NSGS, Azure AD, and so on, implements protection and detections via API logs and alerts.

- **Hot fixes**: If emerging threats are detected in real time, protection can be *patched* into the OAS definition and immediately redeployed.

This shows a very quick overview of how to integrate API monitoring into a SIEM – if you want to explore further, the 42Crunch firewall can be used for free with a trial version of Microsoft Sentinel.

Selecting the correct protections for your APIs

We have covered a number of different shield-right technologies in this chapter, and your ultimate selection will vary according to your budget, technical maturity, skill level, and risk threshold. As a recap, *Table 11.1* shows the technologies' pros and cons.

Solution Type	Pros	Cons
WAFs	Mature and widely available technology, well understood by support teams.	Prone to difficulty in configuring, leading to both high false positives and false negatives.
WAAPs	Promises high accuracy and specificity for API traffic.	Nascent technology with few robust implementations in the marketplace.
API gateways/management	Widely available, with many powerful and flexible features. Almost ubiquitous in all but the smallest organizations.	Require configuration via policies, do not protect against all API vulnerabilities. Can cause performance overheads.

Solution Type	Pros	Cons
API firewalls	Precise enforcement of API contracts.	Requires an accurate contract, additional deployment stage required.

Table 11.1 – Comparison of different shield-right solutions

Summary

In this chapter, we have looked at various technologies that can be used to protect APIs at runtime; this is the shield-right approach to API security. Adding additional protection at runtime can add a further layer of protection to an API that can build upon the strong foundations of a secure design and implementation. Firstly, we looked at some basic techniques to harden the runtime environments of our APIs. WAFs play an important role in protecting web applications and APIs, particularly against some of the older attack methods. We then looked in detail at the vital role that API gateways and API management portals fulfill in securing APIs. In particular, we saw how an open source gateway can provide various protections at runtime, such as rate-limiting, IP address restriction, and JWT validation.

Dedicated API firewalls provide a dedicated layer 7 protection for APIs by enforcing OpenAPI contracts at runtime. Finally, it is important to monitor your API runtimes to identify any anomalies that other runtime protections cannot defend against.

Next, in the final chapter on defending APIs, we will look at how to secure your API microservices.

Further reading

These links provide further reading on hardening Docker and OS images:

- `https://snyk.io/blog/tips-for-hardening-container-image-security-strategy/`
- `https://www.cisecurity.org/insights/blog/how-to-layer-secure-docker-containers-with-hardened-images`
- `https://cheatsheetseries.owasp.org/cheatsheets/Docker_Security_Cheat_Sheet.html`
- `https://www.cisecurity.org/cis-hardened-images/amazon`

These links provide further reading on WAFs:

- `https://www.chakray.com/how-protect-your-apis-installing-configuring-modsecurity-nginx/`

- `https://owasp.org/www-project-modsecurity-core-rule-set/`

- `https://www.netnea.com/cms/apache-tutorial-7_including-modsecurity-core-rules/`

- `https://www.fastly.com/blog/the-waf-efficacy-framework-measuring-the-effectiveness-of-your··waf`

- `https://github.com/coreruleset/coreruleset`

- `https://en.wikipedia.org/wiki/Next-generation_firewall`

- These links provide further reading on API gateways and API management:

- `https://learn.microsoft.com/en-us/azure/api-management/mitigate-owasp-api-threats`

- `https://www.altexsoft.com/blog/api-gateway/`

- `https://konghq.com/install#kong-community`

- `https://www.moesif.com/blog/technical/api-tools/API-Management-vs-API-Gateway-and-where-does-API-Analytics-and-Monitoring-fit/`

- `https://beeceptor.com/`

- `https://curity.io/resources/learn/phantom-token-pattern/`

- These links provide further reading on API firewalls and monitoring:

- `https://docs.42crunch.com/latest/content/concepts/api_firewall.htm`

- `https://42crunch.com/actively-monitor-and-defend-your-apis-with-42crunch-and-the-azure-sentinel-platform/`

- `https://azuremarketplace.microsoft.com/en-us/marketplace/apps/42crunch1580391915541.42crunch_sentinel_solution?tab=Overview&OCID=AIDcmm549zy227_aff_7794_1243925`

12

Securing Microservices

Our penultimate chapter of this book explores how to secure APIs within microservices, an increasingly important topic. Although this topic warrants a chapter of its own, this does not mean that the lessons you have learned about securing API thus far are no longer applicable to microservices. Securing APIs within microservices consists of applying the cornerstone principles we've learned so far, albeit in new and exciting ways.

Firstly, we will understand why microservices arose as a deployment model to understand the motivations for a new architecture better. Then, we'll look at the foundations of microservices and how they can be secured. APIs are all about connectivity, and microservices present unique challenges to securing connections due to the sheer volume of components requiring interconnection in this new architecture. Similarly, access control poses new challenges within a microservices architecture due to the increased client landscape. Finally, we will examine the art of the possible by seeing how a modern microservices stack can be deployed using cloud-native tooling.

By the end of this chapter, you will have a solid foundation of the key elements of building secure APIs in a microservices architecture.

In a nutshell, this chapter is going to cover the following main topics:

- Understanding microservices
- Securing the foundations of microservices
- Securing the connectivity of microservices
- Access control for microservices
- Running secure microservices in the real world

Technical requirements

For this chapter, you will need the following:

- A development machine capable of running Docker locally

- A development machine capable of running VS Code with various marketplace extensions

- Familiarity with Kubernetes and running basic commands

- Access to the internet and a GitHub account to access the examples

This chapter contains sample deployments for various runtime protections, such as Gloo Edge. These configurations and associated instructions will be provided in the GitHub repository for this chapter.

The example code and various breaking changes to the instructions can be found in the `Chapter 12` folder in this book's GitHub repository: `https://github.com/PacktPublishing/Defending-APIs/tree/main/Chapter12`.

Understanding microservices

In this first section, we will gain an understanding of microservices by understanding the issues they intend to address, the unique advantages they offer, and some of the drawbacks of using microservices.

The monolith problem

Before we explore the details of microservices or even seek to define them, let's look at the fundamental problem they sought to solve. Anyone who has worked on large software projects (either as a developer or in an IT operations role) will know the inertia that these projects can develop as they grow and evolve.

These so-called monoliths (generally meaning *a single, large block or structure*) are a conglomerate of all the software components, services, and interfaces required to deliver a business function. Often, a monolith manifests itself as a **three-tier application**, as shown in *Figure 12.1*:

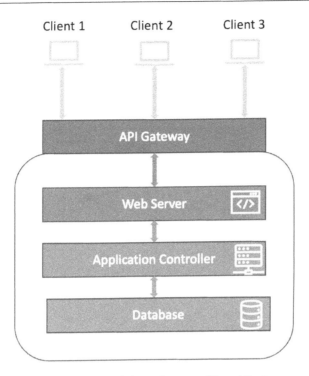

Figure 12.1 – Typical three-tier monolith architecture

Such an architecture requires all components to be released synchronously, meaning that all development teams must align with the same delivery schedule. Additionally, supporting infrastructure (such as API gateways and databases) must also align with this release schedule.

The inevitable downsides of such monolithic development include long release cycles, increased fragility, slow release cycles, inflexibility to embrace new technologies, and high dependency.

From a security perspective, a common problem with a monolith application is the slow release cycle. A penetration test may identify a component with a vulnerability, but due to the tight coupling and interdependency, it is impossible to release a hotfix to only this component. This results in longer windows of exposure to attacks.

Using microservices to build scalable and performant applications

Now that we understand some of the attendant problems of the monolithic application, let's understand the advantages of microservices by looking at a real-world example of microservices in action. The ride-hailing industry is often regarded as a poster child for the success of microservices, so let's consider a fictional service called *Yuba*, as shown in *Figure 12.2*:

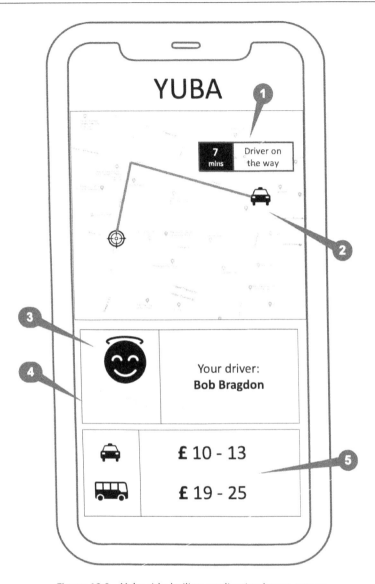

Figure 12.2 – Yuba ride-hailing application home screen

To understand the value of this microservices-based architecture, let's look at each of the identified elements:

- Estimated time of arrival (ETA) of driver
- Driver location tracking via map view
- Driver details, including name and contact details

- Driver rating and feedback
- Trip pricing estimates

Looking at this list of functionalities, it is apparent that each serves a unique and distinct function – the driver's ETA does not depend on the driver's rating, and the location does not depend on the trip price. This is the essence of a **loosely coupled architecture**, where individual components do not depend on one another.

To demonstrate the value of this loosely coupled architecture, let's consider some scenarios. For example, if the driver location tracking team wishes to enhance their service (a higher-precision location), they can redeploy on their microservice without rebuilding the entire application. If the driver rating services were to suffer an outage, the rating would be temporarily unavailable without affecting the rest of the application (this is a graceful failure rather than a catastrophic one). If the trip pricing team wishes to switch to a new backend database, they can do this independently of the rest of the application.

From a secure viewpoint, the most important advantage is that if a vulnerability were discovered in a service (for example, the driver detail widget had a cross-site scripting vulnerability), this service could be disabled while a hotfix is deployed.

While this architecture has numerous advantages, there are some significant drawbacks, including several security considerations.

Risks and challenges of microservices

The most obvious drawback of a microservices architecture is the complexity introduced by a distributed design. With a monolith, it is easy to understand how everything works by simply looking at the code; however, with microservices, we must understand how the different services connect (via APIs) and how they interact. Deploying microservices can also be challenging, particularly in managing their connectivity and dependencies.

The situation is even more complicated from a security perspective. To understand some of the key security challenges, let's look at a simplified microservices application, as shown in *Figure 12.3* (this is a variation of *Figure 11.2*, as described previously in *Chapter 11*):

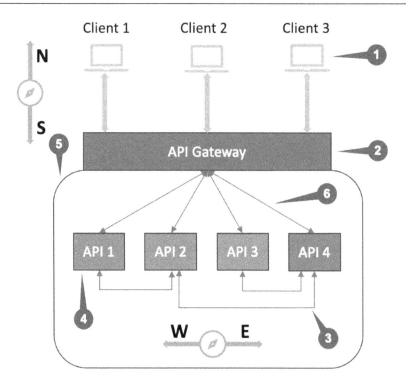

Figure 12.3 – Simplified microservices architecture

This illustration highlights some (but certainly not all) security considerations, as follows:

- **The authentication and authorization of clients and users**: How do the clients and end users authenticate to the gateway, how is this propagated to the backend microservice, and how is authorization managed?

- **Security controls within the API gateway**: The API gateway has a critical role in governing the north-south traffic with a complex myriad of backend microservices. Ensuring that controls are enforced uniformly across all public endpoints is challenging in complex deployments.

- **Intra-service API security**: How are the communications between individual microservices secured? Which microservices are allowed to communicate with one another?

- **API security of the individual microservice**: How does each microservice manage its API security?

- **Security of the microservice perimeter (egress traffic)**: How is external connectivity managed within the microservice perimeter? How is outbound connectivity secured?

- **Inter-service API security (ingress traffic)**: How is the API gateway connectivity to the microservices secured?

This is only scratching at the surface of the challenges of securing APIs in a microservices architecture. Let's dive deeper into these topics, starting with securing the foundations.

Securing the foundations of microservices

Any system is only as strong as its foundations, and nowhere is this more the case than with microservices, which are often built on very disparate foundations. Let's look at how to secure commonly used foundational elements.

Containers

Container security is a core consideration for anyone using containers, particularly in microservices, where they form the core deployment unit.

In the *Container images* section in *Chapter 11*, we looked at the importance of producing secure and hardened container images. If you have not already read that section, now is the perfect time to consider the recommendations there.

The other aspect of using containers securely is to ensure that the containers are being protected at runtime to prevent unauthorized or unexpected container behavior. A container is little more than a process running within a so-called *jail*, preventing it from accessing resources outside of its jail boundary. Unfortunately, there are many ways that a container can be *jailbroken*, allowing processes to access host resources, make outbound connections, or perform other harmful operations.

Fortunately, there are a variety of solutions on the market that provide comprehensive capabilities to monitor container runtime behavior and alert on violations of pre-configured rules. Let's briefly look at the popular CNCF project known as Falco (`https://falco.org`), which offers a community version.

Falco can be installed directly into the host operating system or a Kubernetes cluster. It uses a flexible ruleset to detect various events of interest and can raise alerts to various channels upon triggering. These alerts can be monitored centrally or be used to terminate the affected container.

As a simple example, consider the following rule, which detects the presence of a shell being run within a container. This can be seen as an early indicator of compromise since an attacker may be attempting to gain access to a container and launch attacks via a newly acquired shell.

The highlighted lines show the core rule logic – a shell is spawned in a container other than the specified entry point:

```
- macro: container
  condition: container.id != host
- macro: spawned_process
  condition: evt.type = execve and evt.dir=<
- rule: run_shell_in_container
  desc: a shell was spawned by a non-shell program in a    container.
Container entrypoints are excluded.
```

```
  condition: container and proc.name = bash and spawned_process and
proc.pname exists and not proc.pname in (bash, docker)
  output: "Shell spawned in a container other than entrypoint
(user=%user.name container_id=%container.id container_name=%container.
name shell=%proc.name parent=%proc.pname cmdline=%proc.cmdline)"
  priority: WARNING
```

In this example, the rule will generate an alert with a severity of WARNING to the standard output. Falco also includes a large collection of built-in rules with sensible default values and can be extended using plugins or a gRPC API.

Kubernetes

Although not a prerequisite to building a microservices implementation, the use of Kubernetes as an orchestrator is almost ubiquitous in large microservice implementations. Kubernetes has been one of the core enabling technologies (along with containers) that has made complex deployments a reality by simplifying their management. However, this has come at the cost of greater complexity and a greatly increased attack surface.

In this section, we will look at a few basic recommendations for securing Kubernetes. If you are responsible for administering Kubernetes, consult the in-depth resources in the *Further reading* section.

As a security researcher and curator of an API security newsletter, I report on security issues related to the over-exposure of the Kubernetes API almost weekly. This is by far the most common attack vector for Kubernetes clusters and can be catastrophic if an attacker can gain access, particularly with the default admin role. Be sure to *limit network access to any of the sensitive API ports* on both master and worker nodes.

On a related topic, administrators should also *limit the network access to their clusters* to ensure attackers cannot access the worker nodes or the control plane. All communications between nodes and workers should be secured with **Transport Layer Security** (**TLS**). A key component of Kubernetes is **etcd** (a storage element for state and secrets in the cluster), and as such, *access to etcd should be restricted using strong authentication or network controls*.

Kubernetes offers a comprehensive **role-based access control** (**RBAC**) facility to limit access rights at a fine granularity. Be sure to *use appropriate RBAC controls* in your deployment.

Another common security vulnerability is open access to the Kubernetes administration dashboard, allowing an attacker to control your cluster, similar to them having access to the master API. Be sure to *either disable the administration dashboard completely or implement strict access controls* to minimize the risk of compromise.

We have already looked at best practices for container security. In addition to these, administrators should *restrict the privileges (disable root execution and use read-only filesystems) and resources* available to containers executing on their nodes.

This is a small sample of best practices that should be used; be sure to consult the current recommendations from the official documentation or authoritative sources such as OWASP. With complexity sometimes comes risk.

Observability

The adage *you cannot secure what you cannot see* particularly applies to microservice architectures with their highly distributed nature and a myriad of components and connections. Observing your entire deployment is vital to ensuring its performance and availability. Let's look at the four core elements of observability.

Metrics

Metrics refer to the measurement of various components in the target system. This typically includes CPU usage, memory usage, network throughput and latency, filesystem statistics, API requests, and other performance data. Metrics can usually be expressed as a **count** (the number of events, such as an API request), a **value** (the amount of free memory), or a **histogram** (showing a time sequence of values).

Typical tools for gathering and monitoring metrics include Prometheus, New Relic, Dynatrace, and Datadog.

Tracing

Tracing is important when debugging issues within a system since it allows the support or development teams to be able to trace (hence the name) a sequence of requests, operations, and responses to understand the system's behavior. With appropriate levels of metadata (such as request parameters and response bodies), tracing can allow support teams to recreate failing requests and understand the root cause. Tracing can have significant impacts if overly verbose information is captured.

Some popular tools for implementing tracing in a microservices architecture include Datadog, Honeycomb, Jaeger, Zipkin, New Relic, and Splunk.

Logging

Logging is the most common method of observability and will immediately be familiar to anyone who has ever debugged a computer problem. The system will write a log entry (with a date stamp, severity level, and other metadata and message) to a local or centralized log facility, such as a *syslog* server. Support teams can then analyze the logs to understand a sequence of events to identify unexpected behavior. There is a trade-off with logging: on the one hand, it needs to be sufficiently verbose to provide enough forensics but not so verbose that the volume of logs becomes unwieldy.

Profiling

The final pillar of observability is profiling, which allows support or development teams to gain insights into the system's performance. To do this, various performance data is captured (typically, the duration of an activity) and then presented graphically to show an insight into which system components are performing poorly or introducing bottlenecks. For example, by profiling execution times, the development team can identify which execution paths are consuming most of the time (such as an inefficient database query causing a bottleneck).

Securing the connectivity of microservices

We will now discuss one of the critical aspects of microservices security, namely their connectivity. We will consider both external (so-called north-south traffic) and internal (so-called east-west traffic) connectivity.

Use TLS everywhere

The number one recommendation for ensuring the security of connected devices is to use a secure transport layer, which typically dictates the use of TLS security via the HTTPS protocol.

Client connectivity

Figure 12.3 shows a simplified microservices architecture, where we have identified a security concern with the external clients (labeled **issue 5**). Securing this connectivity is simple with modern API gateways, which have excellent support for HTTP and secure HTTPS client termination. To enable a secure client channel, all that is necessary is to install an appropriate certificate on the API gateway and then ensure that all clients use the secure connection. The security of this channel is reliant on the client verifying the server certificate to detect invalid or expired certificates and to ensure adequate protection is in place to prevent machine-in-the-middle attacks (by using certificate pinning). The **trust relationship is one-way** – the client can verify the provenance of the server, but not vice versa.

Using a real-world example, consider the Kong API gateway we saw in *Chapter 11*, in the *Using API gateways and API management* section. To enable HTTPS connections on Kong, install the **ACME plugin** (named after the protocol used to obtain certificates) and configure a few parameters (such as administrator email); the plugin will handle requesting a certificate from a service such as Let's Encrypt, and then use that to secure all traffic.

Service connectivity

Now, consider the case of securing the connectivity between services; this is depicted via **issue 3** and **issue 6** in *Figure 12.3*. Since both ends of the connection are fixed at deployment time, it is possible to establish a trust relationship between the endpoints using public key cryptography. This allows the use of **mutual TLS (mTLS)**, where both parties verify each other's certificate to ensure that there is a bidirectional trust relationship before communication is permitted. The greatest challenge of mTLS is establishing the initial trust relationship since neither the client nor the server can trust each other, and they must be securely provisioned with public and private key pairs in the form of certificates. To do this, it is necessary to deploy a local **certificate authority** (**CA**) into your microservices environment and, while bootstrapping the cluster, have this server generate and distribute the necessary certificates.

Although this may seem like a significant overhead, there are many benefits of mutual TLS:

- It eliminates machine-in-the-middle attacks
- It prevents spoofing attacks since both identities are guaranteed
- It avoids the dangers of credential-based attacks since stored credentials are not used
- It prevents brute-force attacks since there are no credentials to guess via brute-force
- It prevents malicious API requests since the identity of both the client and the server is guaranteed – the trust is mutual

mTLS should be considered a given in any microservices deployment as it eliminates several attack vectors.

Just enough service mesh

In the introduction to this chapter, I mentioned that we would be looking at some of the state-of-the-art technology that has arisen out of the popularity of microservices as an architecture. In this section, we will briefly examine one of the most exciting technologies to emerge, namely the service mesh.

Service mesh technology is an advanced, complex, and rapidly evolving space; a brief section does not do it justice. Instead, I am going to focus the discussion on the motivation for the service mesh and the problems that it attempts to solve.

Consider a pair of microservices called **API1** and **API2**, as shown in *Figure 12.4*. Since these microservices are developed independently, they happen to have been built on two entirely different technology stacks, namely Spring Boot and FastAPI. While each microservice has a unique business function, they both have a significant common functionality footprint at the network and logging layer. They must both implement similar protocols, handle network outages, perform service discovery, handle retries and failures, and perform failover. Traditionally, it would be the function of the individual development teams to implement all this functionality and then to maintain this over time. The core business logic would take up only a relatively small portion of their development time. The divergent code bases would also open up further attack surfaces since two different code bases and dependent third-party libraries would be required:

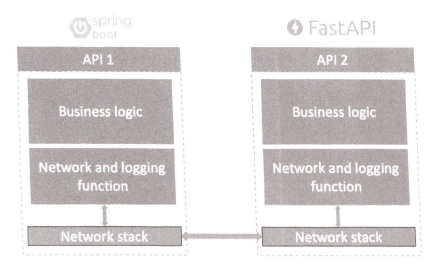

Figure 12.4 – Traditional microservices architecture

Now, let's look at the same two microservices built on top of a service mesh such as **Istio**. *Figure 12.5* shows the same architecture with the following key differences:

- The entire layer of bespoke network and logging functionality has been eliminated

- The services no longer communicate directly over the network

- Service mesh components (typically a reverse proxy such as Envoy) act as the interface between the business logic and the service mesh overlay network:

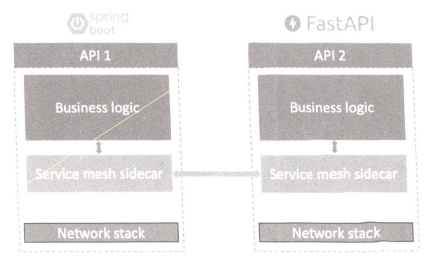

Figure 12.5 – Service mesh-based microservices architecture

The service mesh sidecar (typically Envoy-based) is responsible for handling all the external connectivity-providing features, such as the following:

- **Resiliency of communications**: Retries and timeouts, load balancing, and failover

- **Service discovery**: This enables services to discover one another without having hardcoded dependencies

- **Access control**: Block and allow lists

- **Observability**: Metrics, logging, and tracing of network operations

- **Security**: Via TLS and key management

From a development perspective, the major advantage of the service mesh-based approach is that relatively large swathes of bespoke low-level network code can be eliminated and handed off to a dedicated, well-maintained component.

However, there are significant disadvantages of the service mesh:

- There's a complexity regarding additional components and configurations

- Cost in terms of resources such as memory and CPU

- Early adoption challenges with a relatively nascent technology

The service mesh is becoming increasingly popular, and in the last section of this chapter, we'll look at a real-world example.

Cross-origin resource sharing (CORS)

In *Chapter 10*, we discussed the importance of correctly configuring CORS. For APIs in a microservices environment, the correct application of CORS (via various middleware) is essential to restrict access to a given API. While other controls, such as mTLS and network segregation, can largely prevent an attacker from moving laterally inside a deployment, strict CORS policies add yet another layer of protection.

Access control for microservices

In *Chapter 2*, in the *Access control* section, I mentioned that authentication and authorization are core to the security of APIs, so this book's final section looks at these topics in a microservices environment.

Zero trust at the core

No conversation around access control to microservices could be complete without an overview of the principles of zero-trust computing. In essence, zero trust espouses the principle of *never trust, always verify*. Even if a device or client is on the same network as a server or resource, it should always be fully verified. Never trust a device or client; always fully verify their identity and authorization. This principle should come as no surprise to you after the discussion in preceding sections about always explicitly verifying authentication to avoid broken authentication flaws and explicitly verifying authorization to avoid broken authorization flaws. In a microservices architecture, with its high density of clients and servers and the resultant increased volumes of network transactions, it is even more important to adhere to solid zero-trust principles.

There are seven tenets of zero trust according to NIST SP 800-207 (see the *Further reading* section):

1. All data sources and computing services are considered resources.

2. All communication is secured, regardless of network location.

3. Access to individual enterprise resources is granted on a per-session basis.

4. Access to resources is determined by dynamic policy.

5. The enterprise monitors and measures the integrity and security posture of all owned and associated assets.

6. All resource authentication and authorization is dynamic and strictly enforced before access is allowed.

7. The enterprise collects as much information as possible about the current state of assets, network infrastructure, and communications and uses it to improve its security posture.

The second point was covered in depth in the *Use TLS everywhere* section, while the fifth and seventh points were covered in the *Observability* section. The following sections on authentication, authorization, and token management will focus on addressing the third, fourth, and sixth points.

Authenticate everything

Authenticating the external client (via a web or mobile app) to the API gateway is the most critical authentication challenge, and here, the recommendation is to use OAuth2 with the appropriate flow (see *Chapter 2*, the *OAuth 2.0* section, for a full description).

The OAuth2 termination can be done either in the API gateway itself (for instance, the Kong gateway has a plugin to handle all the OAuth2 flows) or via a separate identity provider, depending on the use case. This second case is shown in *Figure 12.6*:

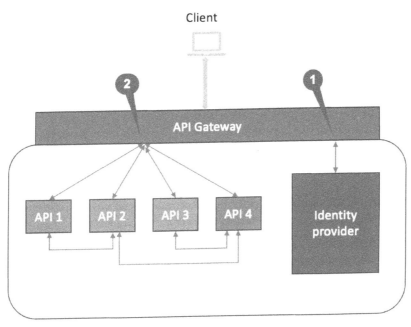

Figure 12.6 – Gateway and separate identity provider

To access the microservice application, the client makes the following calls:

1. A call to the identity provider (routed transparently via the gateway) to authenticate and receive a token (perhaps an OpenID Connect token or a JWT with claims)

2. A call to the API gateway to access the microservice application using the identity or the claims afforded by the token in *Step 1*

Typically, modern architectures will layer OpenID Connect upon OAuth2 to gain a strong identity for the client. **Curity** deems this as the first level of API security maturity (**token-based authentication**) in their maturity model (see the *Further reading* section for further details). At this stage, the server knows the client's identity but has no view of the client's access rights. At this level of maturity, the services behind the gateway will grant access to all clients based on the fact that they present a valid identity. We have successfully validated that they are who they say they are, but we are not validating that they are allowed to do what they are trying to do.

Authorization enforcement

To advance to the next level of maturity (what Curity calls the second level or **token-based authorization**), the system designer must implement some form of authorization server at the API gateway to match an identity to a valid set of authorizations, typically via claims in a token.

The authorization server can either be centralized (this is the simplest approach but it's prone to being a single point of failure) or distributed (the most complex but resilient). Common choices for authorization servers include the **Open Policy Agent** (**OPA**). Claims will be embedded into a JWT, which can be signed and validated by the microservice backend to verify the access rights.

One of the key design decisions with an authorization server topology is whether to use a many-to-one topology (where all microservices communicate with a central authorization server) or a one-to-one topology (where each microservice communicates with its own authorization server).

The many-to-one topology is the simplest to implement and certainly the most commonly used, but it suffers from a few problems:

- The service boundaries are blurred since all authorization decisions are performed in a single central location rather than within the proximity of the service itself

- It opens up the possibility of compromise on the authorization server, which impacts all associated services (effectively, the blast radius is increased)

- The authorization server becomes a single point of failure for all microservices

The one-to-one topology mitigates these disadvantages but at the expense of far greater levels of complexity and configuration.

Token management strategies

The previous discussion illustrated the importance of the token (typically, JWT) in authenticating and authorizing clients in a microservices architecture. These tokens hold a lot of details regarding the client, from their identity to their low-level access rights. As such, it is often desirable to keep the token contents hidden from the public side of the API gateway to prevent external and potentially dangerous parties from gaining access to the raw tokens. It is undesirable to have the client re-authenticate on each transaction, so how can the token be obfuscated on the public interface?

The answer comes via a mechanism known as **opaque tokens** (also known as reference or phantom tokens), where the actual token (containing internal details such as identity and claims) is hidden from the public side by replacing it with an opaque token (which hides the internal details). This exchange operation is performed at the API gateway each time the external client accesses an internal service – the client presents their opaque token, the gateway looks up the actual token, then exchanges this for the actual token, and then forwards the request internally. This method hides the internal details from the public interface. It comes at the disadvantage of some performance overhead (swapping takes processing time) and being reliant on a central single point of failure.

An alternative token-swapping pattern is the so-called **split token** approach, where the JWT is split into two parts: the signature and the header/body pair. The signature is used on the public interface as an identifier, and a cache server at the API gateway extracts the full token from a cache when presented with a signature. The resultant full token can be used for authorization in the internal microservice network. This offers similar advantages to the phantom token approach (hiding token details from the public interface) but removes some of the bottlenecks from the token service.

Running secure microservices in practice

In this chapter's final section, we will look at the current art of the possible when it comes to securing microservices in practice using modern technology stacks based largely on Kubernetes. The previous discussion highlighted modern microservice architectures' increasing complexity in selecting optimum authentication, authorization, and network architectures.

As complexity increases and scale expands, the choices that security engineers face become overwhelming. Several vendors are innovating in this area to ensure a high degree of cohesion of security functionality within microservice architectures so that services are secure by default, or at least security functionality is included as a core consideration rather than an add-on at a later stage.

At the time of writing, some of the vendors in this space are as follows:

- **Traefik Labs** with **Traefik Hub**
- **Ambassador Labs** with **Edge Stack**
- **Nginx** with **Ingress Controller** and **Gateway Fabric** offerings
- **Solo.io** with **Gloo Edge**

I have opted to use the **Gloo Edge** platform for this discussion, but all offerings provide similar functionality and features. In most cases, a highly functional community or trial version will enable you to evaluate functionality for further learning.

Firstly, we'll understand what exactly Gloo Edge offers in terms of architecture, after which we will take a deep dive into some of the key security components. In simple terms, Gloo Edge provides a control plane for the **Envoy** reverse proxy component, allowing the proxy to be controlled in various ways to suit the target architecture. Kubernetes provides an API for an *Ingress Controller,* which allows a reverse proxy (such as Envoy) to provide basic ingress/egress traffic for a cluster. This API is limited to basic network routing, and Gloo Edge leverages Envoy to provide a much more powerful and capable management plane. Simply put, Gloo Edge supercharges Envoy to solve a range of complex ingress/egress challenges such as load balancing, rate limiting, access control, TLS termination, caching, logging, and, of course, security.

Now, let's take a deep dive into the security capabilities offered by Gloo Edge:

- **Cross-site request forgery (CSRF)**: This enforces a CSRF rule in Envoy to ensure appropriate *origin* headers are set in requests and can operate in logging mode or blocking mode.
- **Network encryption**: This manages secure communications via TLS with client connections (outside the perimeter) and mTLS between services within the cluster.

- **Authentication and authorization**: This provides a wealth of different authentication and authorization options covering everything discussed in this book. *Note that many of these features are only available in the Enterprise version.*

- **Rate limiting**: This configures Envoy to provide comprehensive rate-limiting features.

- **Limit active connections**: Similar to the protections offered by rate limiting, limiting active connections can protect the services from being overloaded and failure, and can be handled at the proxy level.

- **Access logging**: This provides extensive support for logging to various targets.

- **Data loss prevention** (**DLP**): This uses regular expression patterns to detect and prevent the egress of sensitive data via an API.

- **Web application firewall**: This provides the capabilities of the **ModSecurity** WAF within the proxy.

- **CORS**: This automates the process of injecting various CORS control headers within the proxy.

- **OPA**: This leverages OPA to prevent further modification of the Gloo Edge configuration.

Given that this is only a list of the security features, this demonstrates the impressive capabilities of modern technology stacks such as Gloo Edge. By leveraging the Envoy proxy, Gloo Edge can use this to enforce a range of security controls, which address many of the top security issues we have discussed in this book. To conclude this section, let's take a deeper dive into one of the more novel protections, namely DLP.

Using the example provided in the Gloo Edge documentation (see the *Further reading* section), let's assume we've submitted the following API request to an echo server (a test API server that returns the parameters as a JSON body):

```
curl $(glooctl proxy url)/ssn/123-45-6789/fakevisa/4397945340344828
```

The response that's received is as follows:

```
{
    "fakevisa": "4397945340344828",
    "ssn": "123-45-6789"
}
```

This example data response contains both a social security number and a Visa card number, which is quite typical of an inadvertent data leakage occurrence. Remember that the OWASP API Top 10 lists *Excessive Data Exposure* as one of the most prevalent API threats.

It is possible to configure the Gloo Edge proxy to explicitly filter these data occurrences from the response using the following configuration:

```
kubectl apply -f - <<EOF
apiVersion: gateway.solo.io/v1
kind: VirtualService
metadata:
  namespace: gloo-system
spec:
  virtualHost:
    domains:
    - '*'
    routes:
    - routeAction:
        single:
          upstream:
            name: json-upstream
            namespace: gloo-system
      options:
        dlp:
          actions:
          - actionType: SSN
          - actionType: ALL_CREDIT_CARDS
EOF
```

This configuration provides an option called `dlp` to the proxy, passing in predefined parameters called `SSN` and `ALL_CREDIT_CARDS` that specify the regular expressions to be applied to filter the response output. Executing the same API command now shows the following response:

```
{
    "fakevisa": "XXXXXXXXXXX4828",
    "ssn": "XXX-XX-X789"
}
```

The API returns the same data; however, this time, the sensitive data is masked by the proxy. For a simple three lines of configuration, we have added a very powerful capability that can transparently protect us from typical data leakage incidents. No changes were necessary in the API code; this new policy could be applied centrally at the edge. In practice, we would want to feed this information back to the development team so that they can implement a fix in the source code. Defense in depth provides the best protection.

I think this is a fine note on which to conclude this chapter. Modern cloud-native implementations such as Kubernetes add great complexity to deployments, potentially increasing the attack surface. Fortunately, vendors are pushing the envelope for highly integrated and capable security capabilities in the fabric of the infrastructure. In the case of our Gloo Edge example, the solution could be deployed with a single Helm command and three lines of configuration – I am optimistic for the future of API security on such platforms.

Summary

By now, you should appreciate the key challenges of securing APIs on microservices. The biggest challenge is the increased complexity of the systems compared to a comparable monolithic application implementation. Due to the increased demands to innovate and deliver functionality, it is safe to say that microservices are here to stay, and security will need to embrace this changing landscape. The recommended approach is to build upon a solid and secure foundation, starting with securing the containers that are used and then the Kubernetes clusters that provide the runtime environments. We learned how to secure the critical connectivity between clients and the gateway using TLS and between the microservices themselves using mTLS. The service mesh promises to secure communications seamlessly while providing integration tracing and monitoring.

Finally, we learned about the foundations of zero trust, which is a key philosophy for ensuring strong access control in a complex, distributed architecture. The patterns for authentication and authorization we saw in earlier chapters can be applied to microservices using the approaches described in this section. Hopefully, you are excited at the possibilities offered by modern, cloud-native solutions such as Gloo Edge in providing a fully integrated, secure platform upon which to build microservices.

We are at the end of the technical section of this book; in the final chapter, we'll look at how to implement an API security strategy.

Further reading

These links provide further reading on understanding the fundamentals of microservices:

- `https://microservices.io/`
- `https://www.openlegacy.com/blog/monolithic-application`

These links provide further reading on securing the foundations of microservices:

- `https://falco.org/`
- `https://www.aquasec.com/cloud-native-academy/kubernetes-in-production/kubernetes-security-best-practices-10-steps-to-securing-k8s/`

- `https://cheatsheetseries.owasp.org/cheatsheets/Kubernetes_Security_Cheat_Sheet.html`

- `https://www.practical-devsecops.com/kubernetes-security-best-practices/`

- `https://arstechnica.com/information-technology/2018/02/tesla-cloud-resources-are-hacked-to-run-cryptocurrency-mining-malware/`

- `https://blog.kubesimplify.com/four-pillars-of-observability-in-kubernetes`

These links provide further reading on securing the connectivity of microservices:

- `https://nvlpubs.nist.gov/nistpubs/SpecialPublications/NIST.SP.800-204A.pdf`

- `https://nvlpubs.nist.gov/nistpubs/SpecialPublications/NIST.SP.800-204.pdf`

- `https://www.cloudflare.com/en-gb/learning/access-management/what-is-mutual-tls/`

- `https://approov.io/blog/how-certificate-pinning-helps-thwart-mobile-mitm-attacks`

- `https://docs.konghq.com/hub/kong-inc/acme/`

- `https://medium.com/microservices-in-practice/service-mesh-for-microservices-2953109a3c9a`

- `https://tetrate.io/blog/mtls-best-practices-for-kubernetes/`

These links provide further reading on access control for microservices:

- `https://curity.io/resources/learn/introduction-identity-and-access-management/`

- `https://curity.io/resources/learn/the-api-security-maturity-model/`

- `https://www.okta.com/resources/whitepaper/8-ways-to-secure-your-microservices-architecture/`

- `https://identityserver4.readthedocs.io/en/latest/topics/reference_tokens.html`

- `https://nvlpubs.nist.gov/nistpubs/SpecialPublications/NIST.SP.800-207.pdf`

- https://www.cisa.gov/sites/default/files/2023-04/zero_trust_maturity_model_v2_508.pdf

These links provide further reading on running microservices in practice:

- https://docs.solo.io/gloo-edge/latest/
- https://traefik.io/traefik-hub/
- https://docs.nginx.com/nginx/
- https://www.getambassador.io/products/edge-stack/api-gateway
- https://traefik.io/blog/reverse-proxy-vs-ingress-controller-vs-api-gateway/
- https://docs.solo.io/gloo-edge/latest/introduction/architecture/deployment_arch/

13
Implementing an API Security Strategy

This book's final chapter focuses on applying the knowledge you've gained over the last 12 chapters to create a comprehensive API security strategy for your organization. Your strategy will depend on your current position and your security goals and will largely be influenced by your organizational structure, particularly who owns APIs and their security. In the first section, we will examine typical organizational stakeholders and how their roles and responsibilities must be aligned as part of your strategy. We will then examine the 42Crunch API security maturity model to understand the six domains of API security and what maturation looks like for each domain. After, we'll dive into rolling out your API security strategy, firstly by planning your objectives against the current state and capabilities and, secondly, by running your strategy as part of your daily process. Finally, we'll conclude this chapter – and this book – by looking at your ongoing personal API security journey.

In a nutshell, this chapter is going to cover the following main topics:

- Ownership of API security
- The 42Crunch maturity model
- Planning your program
- Running your program
- Your personal API security journey

Ownership of API security

Your API security strategy cannot exist in isolation from your organization's API business and development strategy. As a security leader, you must understand the other stakeholders responsible for API strategy and delivery to ensure that your security strategy aligns with their objectives.

Ownership of APIs tends to vary from one organization to another, and there is no hard and fast rule regarding how it is assigned. In the *Further reading* section, there is a reference to a blog post from MuleSoft that describes a typical pattern for API ownership; we will use this to frame our discussion. This is shown visually in *Figure 13.1*:

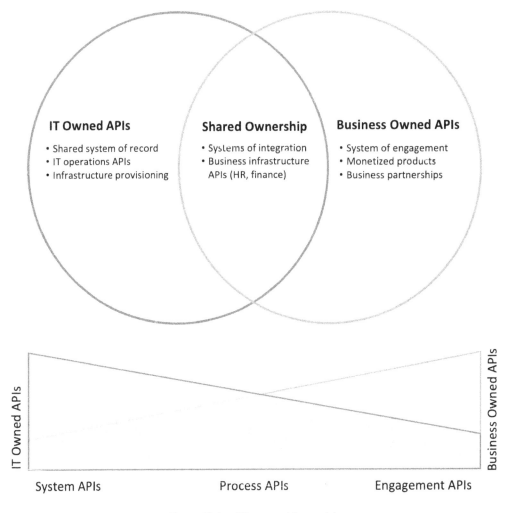

Figure 13.1 – API ownership model

There are three main API owners in this model:

- **IT-owned APIs**: This ownership model aligns most closely with traditional IT systems where the IT department wholly owns the resources. These are core services such as infrastructure provisioning or operations such as ServiceNow or Salesforce platforms. These are often called **system-of-record** APIs, which favor stability and accuracy over features and functionality.

- **Business-owned APIs**: This ownership model reflects the rise of APIs-as-a-product organizations (a typical example is Twilio, where the API is their product) that derive their business value through APIs, which are their **system of engagement**. Here, innovation and agility are key in monetizing existing services to new markets or in delivering new business partnerships. The APIs are owned by the business units, together with their development teams.

- **Shared ownership APIs**: Sitting in the middle ground between these two models is the shared ownership model, which is mostly used to integrate systems such as HR, finance, or inventory systems. The IT department will still exert some control over the operation of these APIs, but their ultimate ownership may reside with other organizational departments.

It is important to understand the ownership of APIs when it comes to kicking off an API security initiative since the ownership will determine the business risk appetite. For **systems of record**, where data integrity is paramount, the owners will likely have a low-risk appetite and engage fully with an API security initiative. They will appreciate the value that such an initiative brings toward producing more dependable APIs that match their objectives. Additionally, many IT security teams may report to the head of IT and have a common senior manager, which will help with the adoption of such an initiative.

The situation is different in the case of the business-owned APIs functioning as a **system of engagement**. Here, the focus is on innovation and agility and an API security initiative will possibly be regarded as an inhibitor to rapid innovation or short release cycles. The risk appetite will likely be higher in this case; in many cases, the business is doing a quick proof of concept that's limited in scope and is less concerned about the consequences of lax security. If this is the case, you will have additional challenges in gaining traction with your initiative. The key recommendation is to form a working relationship with the business unit and reassure them that you are not there to slow them down but rather to aid as an accelerant for innovation while also improving the security posture of their portfolio. These two objectives are not mutually exclusive, given a collaborative working arrangement.

Understanding your stakeholders

Now that we understand the different ownership models, let's look at the various personas across the IT, API, operations, security, and business units to understand their perspectives on API security.

Table 13.1 shows the typical roles in the security domain, along with their key responsibilities:

Role	Description
CISO	They are responsible for information security in an organization
Head of AppSec	They are responsible for the AppSec program and activities in the organization
DevSecOps team	This team is responsible for the integration and operation of security tools within the automated SDLC environment

Role	Description
Pentest/red team	This team is responsible for offensive testing of product releases using black box techniques
Risk and compliance team	This team is responsible for managing risk and compliance in the organization based on applicable operating environments

Table 13.1 – Typical roles in the security domain

Table 13.2 shows the typical roles in the business or development domain, along with their key responsibilities:

Role	Description
CIO	They are responsible for the IT operations of a business unit
Product owner	They are responsible for the product management of a business unit's offerings
Technical lead	They are responsible for managing the technical team on a given product
Solution architect	They are responsible for supporting and evangelizing the product to the customer
DevOps team	This team is responsible for the operation of the build and release process through automation

Table 13.2 – Typical roles in the business or development domain

Table 13.3 shows the typical roles in the API product domain, along with their key responsibilities:

Role	Description
API product owner	They are responsible for the product management of a set of APIs offered as a product
API platform owner	They are the owner of the central API platforms (API gateways and management portals) and API PaaS infrastructure
API architect	They are responsible for the overall API strategy (authentication, authorization, and architecture) in the organization

Table 13.3 – Typical roles in the API product domain

The reason for deliberately and explicitly enumerating this array of stakeholders is to illustrate that ownership of API security potentially resides with several organizational units. For example, the API platform owner may be responsible for applying appropriate API gateway policies, the API architect may be responsible for the overall authentication strategy, the head of AppSec may be responsible for SAST/DAST scans, and the CISO is ultimately responsible for the security of the API.

To avoid duplication of responsibilities (or worse still, neglect of responsibility), the roles and responsibilities should be clearly defined.

Roles and responsibilities

In *Chapter 1, What Is API Security?*, we saw the typical DevOps cycle in *Figure 1.3*. The cycle has eight SDLC phases: plan, code, build, test, release, deploy, operate, and monitor. In an ideal world, security touchpoints across all eight stages of the SDLC will be encapsulated perfectly by the "shift-left, shield-right" mantra used in this book.

Unfortunately, in reality, many organizations have relatively immature security processes typically characterized by late-stage monitoring of IT systems via an SIEM and SOC. This monitoring typically only identifies attacks or threats as they occur and is a reactive security practice. This monitoring is usually performed by the IT security team, with the CISO being accountable, and is shown in the SDLC in *Figure 13.2*:

Figure 13.2 – Monitoring in the SDLC

With the growing awareness of application security as a necessary pre-emptive control, more mature organizations have sought to introduce SAST/DAST scanning into the release and deploy cycle. These tools are usually integrated with the help of the DevSecOps team, and the CISO is still accountable for managing the risk identified in this process, as shown in the SDLC in *Figure 13.3*:

Figure 13.3 – Monitoring and scanning in the SDLC

The benefits of shifting security left, or earlier in the SDLC, have been widely accepted in the industry over the last decade. Typically, this involves co-opting development teams into the security conversation as they take measures to ensure that they are producing secure code by using secure libraries, understanding risks and threats, using secure coding patterns, testing for secure vulnerabilities in the development phases, and more. Furthermore, the accountability for security is shifted from the CISO to the business or product owners, who increasingly regard a secure product as a differentiating feature in the marketplace. An example shift-left SDLC is shown in *Figure 13.4*:

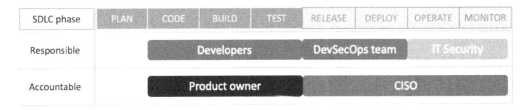

Figure 13.4 – Shift-left in the SDLC

While this shift-left pattern makes great steps toward producing more secure code, it can be improved further by incorporating security right at the design stage to ensure security is considered part of the design before any code is written. As an example, in an API security context, this may relate to decisions about managing identities and roles within an API product. A fully integrated secure SDLC is shown in *Figure 13.5*:

Figure 13.5 – Fully integrated secure SDLC

These examples illustrate that the ownership of both APIs and API security is a complex topic and that often, this ownership will be distributed across two or more organizational units. A distributed ownership model has benefits since it allows domain experts to focus on their specialism: the CISO ensures the APIs are monitored optimally in the SIEM and SOC, and the product owner ensures that the APIs are well-architected and secure by design.

The 42Crunch maturity model

In my time as a technical evangelist at 42Crunch, I formulated a six-domain API security maturity model that has proved to be popular with customs in determining both their current security posture and their roadmap toward a more secure posture.

The maturity model features a set of activities for each domain, which may exist to varying degrees based on maturity. For this discussion, we will bucket the activities as **non-existent**, **emerging**, or **established**.

Inventory

An up-to-date and accurate inventory is key to maintaining visibility into the exposed risk and attack surface.

The adage "*you can't protect what you can't see*" applies perfectly to API security. As APIs grow exponentially, fueled by business demand, it is increasingly difficult for security teams to maintain visibility of what APIs exist and what risks they expose.

Three elements are key:

- How new APIs are introduced and tracked in the organization
- Discovering the API inventory by introspecting the source code repositories to discover hidden API artifacts
- Runtime discovery of APIs (via network traffic inspection, and so on)

At the lowest maturity level, only a basic inventory (usually, APIs deemed critical for the business) is maintained via spreadsheet or manual tracking. There is no management of shadow/zombie APIs.

At the emerging maturity level, an inventory is maintained via API management or a centralized platform. A standard process is used for new API development. For an established level of maturity, the inventory is actively tracked via a centralized platform, and shadow and zombie APIs are deprecated and upgraded.

Design

It is significantly more cost-effective to address security issues at the design phase rather than later in the life cycle – a shift-left approach is key.

A solid API design practice is the foundation of usable, scalable, documented, and secure APIs. Here are some of the key elements of secure API design:

- Authentication methods
- Authorization models and access control

- Data privacy requirements

- Data exposure (explicit/least content/fit for purpose) requirements

- Compliance requirements

- Account reset mechanisms

- Use and abuse cases

- Key and token issue and revocation methods

- Rate limiting and quota enforcement

Additionally, API design teams should perform threat modeling exercises to understand their threat environment and attack surface.

At the lowest maturity level, no formal API design process is in place; instead, a code-first method is used. There is usually no upfront consideration of security concerns, threats, compliance, and data privacy.

At the emerging maturity level, APIs are developed using a design-first approach based on OAS definitions. Security concerns are addressed on an ad hoc basis with no standard process. For an established level of maturity, security is a first-class element of API design that includes standard patterns/practices such as threat modeling.

Development

This vital stage is where the rubber meets the road – developers should ensure they follow security best practices to avoid introducing vulnerabilities into APIs.

A crucial element of secure APIs is the development process, where specifications are implemented in live APIs. Some key considerations here are as follows:

- Choice of languages, libraries, and frameworks

- Correct configuration of frameworks to ensure security best practices are followed

- Defensive coding – do not trust user input and handle all unexpected failures

- Use central points of enforcement of authentication and authorization – avoid "spaghetti code"

- Think like an attacker!

At the lowest maturity level, developers are largely unaware of security concerns in API development or approaches to secure code in general.

At the emerging maturity level, developers are familiar with security considerations and use secure coding practices, albeit sporadically. For an established level of maturity, developers are fully versed in secure code and API security topics and proactively seek to use best practices and defensive coding.

Testing

Without adequate API security testing, an organization runs the risk of deploying insecure APIs – test early, test often, test everywhere.

API security testing is vital to ensure that APIs are verified as secure before deployment. Security testing should be tightly integrated into the CI/CD process and should avoid any manual effort. Tests should be able to "break the build" in the event of failure. The following aspects should be tested:

- Authentication and authorization bypass
- Excessive data or information exposure
- Handling invalid request data correctly
- Verifying response codes for success and failures
- Implementation of rate-limiting and quotas
- Changes in configuration in production environments from their desired target state (so-called configuration drift)

At the lowest maturity level, there is no specific API security testing, with only functional testing in place.

At the emerging maturity level, API security testing largely uses manual testing and lacks automation and CI/CD integration. For an established level of maturity, API security testing is tightly integrated into all stages of the SDLC, and failures can block releases.

Protection

A defense-in-depth approach is the foundation of risk reduction – regardless of how well-designed your APIs are, they will still be attacked by persistent and skilled adversaries.

Despite the best efforts during the preceding phases of the SDLC, APIs will still come under attack and should be protected via dedicated API protection mechanisms.

API protection should include the following:

- JWT validation
- Secure transport options
- Brute-force protection
- Invalid path or operation access
- Rejection of invalid request data
- Filtering of response data

Protection logs should be ingested into standard SIEM/SOC platforms to ensure visibility of API security operations.

At the lowest maturity level, no specific API runtime protection is implemented; standard firewalls or WAFs are the only protection in place.

At the emerging maturity level, some protection is provided, typically using API gateways to provide basic enforcement of rate-limiting, token validation, and more. For an established level of maturity, dedicated API firewalls are implemented to provide localized protection at the API transaction level.

Governance

Trust but verify – a robust governance process is essential to ensure that API development observes organizational methodologies.

The final domain of API security is the overall governance process, which ensures that APIs are designed, developed, tested, and protected according to the organization's process.

Governance covers the following principles:

- APIs are consistent – that is, they use standard patterns for authentication and authorization
- Standard processes, including testing and remediation requirements, are followed to develop new and updated APIs
- Data privacy and compliance requirements are met
- A process is observed for APIs at their end of life to eliminate insecure zombie APIs
- Stakeholders are enabled on API-specific security topics
- Enablement is updated based on emerging threats
- API development is largely ungoverned, with business units each using their own process with no central oversight

At the lowest maturity level, governance addresses only the basic requirements of compliance and regulatory requirements.

At the emerging maturity level, governance is proactive, and APIs are developed to a standard process. For an established level of maturity, deviations are tracked, and discrepancies are addressed.

Planning your program

Now that you have examined the key topics of API and API security ownership and have the foundations of a maturity model, it is time for the rubber to hit the road as you begin to plan your program.

Establishing your objectives

Simon Sinek's seminal TED talk *Start with Why* inspires leaders and organizations to understand their motivation for what they do and the importance of the "why" they do what they do. The same can be said for establishing an API security program – without clear objectives or *raison d'être*, your program may flounder and fail. You need to understand the compelling reason(s) for implementing a change program of scale. Perhaps you process medical records and cannot risk an API breach disclosing patient data. Or maybe you are a payment processor that is bound by strict regulatory requirements. Or perhaps you are an "API-first" company whose very business succeeds (or fails) on the strength of their APIs and their security.

Find your why, use that to determine your requirements, and plan your program accordingly.

One of the biggest blockers to starting a software security initiative is feeling overwhelmed when establishing the first steps on the journey, particularly if you are responsible for a large estate. With so many applications or APIs, where do we even start? Unfortunately, this hesitancy can undo many initiatives since they fail to even get started, or if they do, the effort is wasted on lesser important assets.

In *Chapter 1, What Is API Security?*, I introduced a basic approach to a *risk-based methodology* to prioritize the most important APIs based on your business's security objectives. Use this method to identify your highest priority cohort of APIs, enroll them into your program, and then spread the selection criteria wider to the next cohort and then the next. Avoid the temptation to take on your entire portfolio in one fell swoop as this can lead to early failure when no progress is apparent.

Assessing your current state

You may be working in a greenfield environment regarding API security where nothing is in place, and you are building from ground zero. Or – more likely – you have a brownfield environment where some API security measures are already in place, and you are seeking to improve or mature these. In this case, you must get a good estimate of your current state before making changes.

The first task is to estimate your inventory. This can be done in several ways:

- Scan your Git repositories for identifiers that indicate API code (OAS definitions, route controllers in source code, and more)

- Ingest network traffic and use tooling to identify API-specific traffic

- If you are running an API gateway, extract their inventory of proxies and use this as a starting point

Once you have identified the APIs, use your organizational knowledge to determine ownership of the API, and then use discovery meetings to map their current capabilities to a maturity model (such as the 42Crunch model or more general-purpose models such as **OpenSAMM** or the **NIST SSDF**).

Lastly, decide what controls, processes, and procedures are required to attain your security requirements. Build consensus and awareness with all stakeholders and bake in review, updates, and enablement to ensure longevity and relevance.

Building a landing zone for APIs

In recent years, one of the more useful paradigms that has emerged from the DevSecOps movement is the concept of so-called guardrails or landing zones. The concept is simple: build a zone (comprising elements from across the eight phases of the SDLC) with security tooling baked into the fabric of the zone. The only thing the application development team needs to do is contribute their code; the rest of their environment is already configured optimally for their development process and, of course, for secure development and deployment.

These landing zones can become unwieldy for large, complex, heterogeneous applications requiring many languages, toolkits, SDKs, environments, and so on. However, for APIs, the story is a little simpler. An organization can realistically standardize one or two landing zones and coax development teams to use these secure landing zones to achieve greater API security. It becomes a compelling proposition for security and development teams alike:

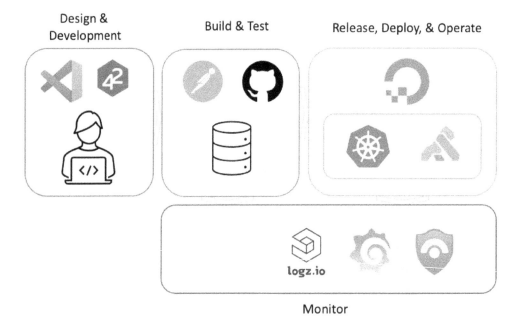

Figure 13.6 – Sample secure API development landing zone

In the sample secure API development landing zone shown in *Figure 13.6*, the following key components are provided:

- A design and development zone based around a VS Code editing experience, with 42Crunch providing OAS definition auditing and instantaneous API testing in the editor

- A build and test zone based on GitHub and its Actions component for automation, and Postman to run automated API testing using a **command-line interface** (**CLI**)

- A release, deploy, and operate zone based on a DigitalOcean tenant (this could be any similar cloud provider) with Kubernetes as a deployment orchestrator using a Kong API gateway as the main ingress from the outside world

- A monitoring zone that uses Azure Sentinel as the SIEM, Grafana as the telemetry and instrumentation dashboard, and `Logz.io` as the logging facility

This is purely an illustrative example; the specific details will vary according to your organization. The key takeaway is that you build out a landing zone with the required elements and then secure that by providing secure defaults and hardened configurations. A developer should be able to go from code to a secured API without having to concern themselves with any intermediate steps.

Running your program

Once you have established your program's goals and identified your stakeholders, you can start running your program. To do that, first and foremost, you need a team composed of the right people for the job. The trick is to find the right people; let's look at some approaches.

Building your teams

First up, you need to build your own team who will work to achieve your objectives. Adam Shostack has written an excellent blog on the topic (see *Further reading*), and his perspectives reflect my reality of having built several large-scale AppSec programs. The key point is the hardest one to grasp: to build an AppSec team, you do not need a team of AppSec specialists. Shostack expresses it perfectly: "*by using exceptional talents doing over-specialization.*" While securing software has an obvious technical element to it, by far, the biggest challenges are human-centric. You will, above all else, require the buy-in and cooperation of your development teams; after all, it is these teams who need to make changes to their processes or fix their code to improve security. What is needed most are diplomats who can lead with empathy and negotiate change. Deep technical specialists are often too inclined to want to solve the problem themselves rather than empowering others to solve their problems. Teams of deep technical specialists will not scale.

Speaking anecdotally from my experience (see the YouTube talk in *Further reading*), I assumed I needed both InfoSec skills (with qualifications such as CISSP and others) and pentest skills when building my team in my first role. Instead, I was assigned a team of generalists with very diverse skills and backgrounds, and we shaped this into a very successful AppSec team.

The other way to approach the scaling issue with AppSec teams is to delegate or outsource much of the day-to-day operation to security-minded members of the application development teams themselves. This is the now popular security **champions approach**, where the champions are responsible for activities such as evangelizing security, developing security standards and policies, running events and activities, doing threat modeling, performing code reviews, and using security testing tools. Think of the security champion as a local extension of your team. Clarity is essential to determine who has ultimate accountability for security decisions. OWASP offers excellent guidance on this topic (see *Further reading*).

Tracking your progress

As your program proceeds, you will need to demonstrate progress. This is best done by selecting a variety of **key performance indicators** (**KPIs**) that reflect the objectives of your program.

Understanding your KPIs

For each metric chosen, several metadata properties indicate the metric's nature:

- **Data source**: This indicates how the metric is measured. Possible values include the 42Crunch platform, platform metrics (GitHub, CI/CD, and so on), SIEM/SOC, the API gateway, and surveys, which are the least accurate as they tend to be subjective.

- **Classification**: This indicates whether the value should increase or decrease to reflect an improvement. Metrics with a decreasing value are good since a value of zero indicates no further improvement is needed, or metrics measured as a percentage with an increasing value are good since 100% indicates no further improvement. For open-ended metrics, a baseline should be established based on the organization's goals and objectives.

- **Unit of measurement**: The units of measurement can be one of the following: a count (a numeric value), percentage, duration (a time or date value), or a Boolean (a yes/no value).

- **Trend**: This specifies whether the metric is increasing or decreasing to indicate improvement. Flaw count is a good example of a decreasing metric, while code coverage is a good example of an increasing metric.

- **Reliability**: This indicates the approximate reliability of the metric. Some metrics that are measured from platforms (42Crunch, GitHub, and so on) can be precise with high reliability, while others can be less reliable, such as detected threats. Others are entirely subjective, such as the cost to remediate.

- **Leading or trailing**: This indicates whether the metric shows the impact of past behaviors (trailing) or that future KPIs should improve based on the measurement based on correlation (leading). Education, enablement, and engagement are examples of leading indicators. Scan results and bug bounty reports are examples of trailing indicators.

Ensure that you understand the nature of your KPIs to determine sensible target values and timescales as these will likely be necessary to achieve them.

Selecting your KPIs

As part of the research in my role as a technical evangelist, I have collated nearly 200 KPIs for API security and made these available in spreadsheet form in this book's GitHub repository.

The KPIs are sorted according to the phase of the SDLC. The typical KPI types are as follows:

- **Plan**: Security definitions metrics, transport security metrics, input validation and parameter security metrics, output and response security metrics, error handling and information disclosure metrics, compliance metrics, and threat modeling metrics

- **Code**: Pull request review metrics, pull request integration metrics, code review metrics, code scanning metrics, and dynamic testing metrics

- **Build**: Dependency and component security, tooling and automation metrics, development, and remediation efficiency metrics

- **Test**: API security testing

- **Operate**: Vulnerability metrics, threat intelligence metrics, performance metrics, and security incident metrics

- **Monitor**: Security monitoring metrics, traffic metrics, and cost metrics

To start with, pick a half dozen or so of the key metrics (and possibly ones you know you can improve), use these as a yardstick for your program, and then expand as your program gathers momentum. From my experience, I would pick metrics based on the measurement of authentication and authorization in the development phase and then coverage metrics for testing.

Integrating with your existing AppSec program

In the *Roles and responsibilities* section, we discussed how most organizations will likely have some form of DevSecOps team performing security testing and scanning activities (see *Figure 13.3*). As you build out an API security initiative, aligning your efforts and activities with those of the DevSecOps team is wise.

Integrate API testing methods

For each of the common security testing types, particular touchpoints applicable to API security are worthy of focus, either by creating specific rules, adjusting the severity categories, or providing remediation advice. For example, here are some suggestions:

- **Static application security testing** (**SAST**): For SAST scans, pay particular attention to issues that have been identified relating to injection vulnerabilities since APIs are prone to injection attacks via external payloads. Also, be aware of findings related to data processing, such as **XML external entity** (**XXE**) based attacks.

- **Dynamic application security testing** (**DAST**): For DAST scans, identify endpoints that are not configured correctly with TLS or fail under high load conditions or with large payloads. These endpoints may be fragile and susceptible to denial-of-service attacks and should be hardened or protected with rate-limiting solutions.

- **Software composition analysis** (**SCA**): For SCA testing, identify packages or components critical to APIs (frameworks such as **Connexion** or **FastAPI** in a Python application, for example) and ensure that the versions that are used are not affected by open vulnerabilities.

Work closely with your AppSec team to leverage their efforts and, in particular, learn from their hard-learned lessons.

Understand your API dependencies

We have already discussed the importance of managing and maintaining your software dependencies to ensure that you are not inheriting risk from vulnerable software. It is also important to understand the provenance of your upstream APIs:

- Do you know if these APIs are built using secure software development methodologies?

- Do you know if these APIs are tested for vulnerabilities?

- In the event of vulnerabilities being discovered, do you have an agreement in place with your supplier to remediate these vulnerabilities?

It is highly advisable to map out a dependency tree of your API infrastructure to identify all contributing elements, including software components or third-party APIs. Use this dependency map to build out a view of your risk profile.

Your personal API security journey

We are now at the end of this book, but that does not mean that your personal API security journey has concluded. I would like to think it has only just started. APIs and API security are rapidly evolving domains, with new technologies (such as GraphQL) posing new risks to organizations. Hopefully, this book has given you a solid foundation in the basics of API security, how to attack APIs, and, most importantly, how to defend them.

To keep up to date on all breaking news relating to API security, including breaches, views and opinions, tools, and techniques, I would recommend the bi-weekly newsletter I curate at APISecurity.io (`https://apisecurity.io/`).

If you prefer a more tactile, hands-on approach to learning, then the good folks at APISecurity University have several online training courses on various API security topics (`https://www.apisecuniversity.com/`).

Happy learning!

Summary

This brief chapter covered the very important topic of building an API security strategy and saw the theory we have learned about API security applied to real-world API development. Understanding who owns your APIs is important in understanding how to drive the messaging around the need for API security. A broad-based approach involving the CISO or IT security organization and their colleagues in the API product development teams is likely to produce the best results since this will include API security touchpoints across all phases of the SDLC.

First, we learned how to plan an API security initiative by understanding our objectives (the "why") and then understanding our current state to form our strategy. We then looked at running a program, focusing on the critical step of building our team and selecting our KPIs to gauge our progress.

Finally, your own continued learning is important for staying on top of emerging threats and changes in technology landscapes.

This chapter brings our journey together to a conclusion. In hindsight, we have covered a lot more material than I might have anticipated when I started with the first chapter nearly 18 months ago. The journey has taken us from the very fundamentals of API requests using HTTP, through to a solid understanding of the core building blocks and then the top vulnerabilities affecting APIs. We then learned how to think like an attacker in understanding how our APIs could be attacked.

The last part of this book honed in on the key topic of defending APIs and took us on a journey of methods and techniques to use throughout the SDLC. A good API security strategy covers touch points from the start of the SDLC (the so-called shift-left approach) to the API runtime (the so-called shield-right approach). Hopefully, at this point, you are inspired and motivated to start securing your organization's APIs.

As mentioned previously, your learning journey is only just beginning, so please do keep in touch, either with me or via this book's dedicated resources, namely the GitHub repository or the YouTube "Code in Action" channel. As I conclude this book, I am happy to report the great success of my first full-day workshop accompanying this book – be sure to keep an eye open for future workshops coming to a venue near you.

Further reading

These links provide further reading on the ownership of API security:

- `https://blogs.mulesoft.com/api-integration/strategy/api-ownership-enterprise/`
- `https://medium.com/apis-and-digital-transformation/without-an-api-product-owner-your-apis-have-a-limited-lifespan-6df98d6ad281`
- `https://www.traceable.ai/blog-post/what-is-api-ownership`

These links provide further reading on the 42Crunch API security maturity model:

- `https://42crunch.com/ebook-api-security-blueprint/`

These links provide further reading on planning your program:

- `https://www.ted.com/talks/simon_sinek_how_great_leaders_inspire_action?language=en`
- `https://csrc.nist.gov/pubs/sp/800/218/final`
- `https://www.opensamm.org/`
- `https://dzone.com/articles/how-to-bring-the-power-of-security-guardrails-to-y`

These links provide further reading on running your program:

- `https://shostack.org/blog/application-security-team/`
- `https://owasp.org/www-project-security-culture/v10/4-Security_Champions/`
- `https://portswigger.net/web-security/xxe`
- `https://www.youtube.com/watch?v=gEDjtsgun1k`

These links provide further reading on your personal API security journey:

- `https://apisecurity.io/`
- `https://www.apisecuniversity.com/`

Index

Symbols

A

`Packtpub.com`

Subscribe to our online digital library for full access to over 7,000 books and videos, as well as industry leading tools to help you plan your personal development and advance your career. For more information, please visit our website.

Why subscribe?

- Spend less time learning and more time coding with practical eBooks and Videos from over 4,000 industry professionals
- Improve your learning with Skill Plans built especially for you
- Get a free eBook or video every month
- Fully searchable for easy access to vital information
- Copy and paste, print, and bookmark content

Did you know that Packt offers eBook versions of every book published, with PDF and ePub files available? You can upgrade to the eBook version at `packtpub.com` and as a print book customer, you are entitled to a discount on the eBook copy. Get in touch with us at `customercare@packtpub.com` for more details.

At `www.packtpub.com`, you can also read a collection of free technical articles, sign up for a range of free newsletters, and receive exclusive discounts and offers on Packt books and eBooks.

Other Books You May Enjoy

If you enjoyed this book, you may be interested in these other books by Packt:

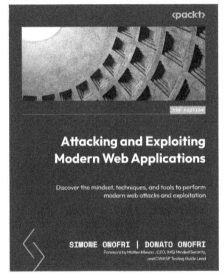

Attacking and Exploiting Modern Web Applications

Simone Onofri, Donato Onofri

ISBN: 978-1-80181-629-8

- Understand the mindset, methodologies, and toolset needed to carry out web attacks
- Discover how SAML and SSO work and study their vulnerabilities
- Get to grips with WordPress and learn how to exploit SQL injection
- Find out how IoT devices work and exploit command injection
- Familiarize yourself with ElectronJS applications and transform an XSS to an RCE
- Discover how to audit Solidity's Ethereum smart contracts
- Get the hang of decompiling, debugging, and instrumenting web applications

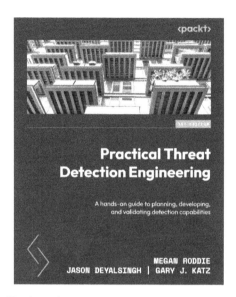

Practical Threat Detection Engineering

Megan Roddie, Jason Deyalsingh, Gary J. Katz

ISBN: 978-1-80107-671-5

- Understand the detection engineering process
- Build a detection engineering test lab
- Learn how to maintain detections as code
- Understand how threat intelligence can be used to drive detection development
- Prove the effectiveness of detection capabilities to business leadership
- Learn how to limit attackers' ability to inflict damage by detecting any malicious activity early

Packt is searching for authors like you

If you're interested in becoming an author for Packt, please visit `authors.packtpub.com` and apply today. We have worked with thousands of developers and tech professionals, just like you, to help them share their insight with the global tech community. You can make a general application, apply for a specific hot topic that we are recruiting an author for, or submit your own idea.

Share Your Thoughts

Now you've finished *Defending APIs,* we'd love to hear your thoughts! Scan the QR code below to go straight to the Amazon review page for this book and share your feedback or leave a review on the site that you purchased it from.

`https://packt.link/r/1804617121`

Your review is important to us and the tech community and will help us make sure we're delivering excellent quality content.

Download a free PDF copy of this book

Thanks for purchasing this book!

Do you like to read on the go but are unable to carry your print books everywhere?

Is your eBook purchase not compatible with the device of your choice?

Don't worry, now with every Packt book you get a DRM-free PDF version of that book at no cost.

Read anywhere, any place, on any device. Search, copy, and paste code from your favorite technical books directly into your application.

The perks don't stop there, you can get exclusive access to discounts, newsletters, and great free content in your inbox daily

Follow these simple steps to get the benefits:

1. Scan the QR code or visit the link below

https://packt.link/free-ebook/9781804617120

2. Submit your proof of purchase

3. That's it! We'll send your free PDF and other benefits to your email directly

www.ingramcontent.com/pod-product-compliance
Lightning Source LLC
Chambersburg PA
CBHW080610060326

40690CB00021B/4648